STEROID CONTRACEPTIVES AND WOMEN'S RESPONSE

Regional Variability in Side-Effects and Pharmacokinetics

REPRODUCTIVE BIOLOGY

Series Editor: Sheldon J. Segal

The Population Council
New York, New York

Current Volumes in this Series

AIDS AND WOMEN'S REPRODUCTIVE HEALTH
Edited by Lincoln C. Chen, Jaime Sepulveda Amor, and Sheldon J. Segal

AUTOCRINE AND PARACRINE MECHANISMS IN REPRODUCTIVE ENDOCRINOLOGY
Edited by Lewis C. Krey, Bela J. Gulyas, and John A. McCracken

CONTRACEPTIVE STEROIDS: Pharmacology and Safety
Edited by A. T. Gregoire and Richard P. Blye

DEMOGRAPHIC AND PROGRAMMATIC CONSEQUENCES OF CONTRACEPTIVE INNOVATIONS
Edited by Sheldon J. Segal, Amy O. Tsui, and Susan M. Rogers

ENDOCRINE AND BIOCHEMICAL DEVELOPMENT OF THE FETUS AND NEONATE
Edited by José M. Cuezva, Ana M. Pascual-Leone, and Mulchand S. Patel

GENETIC MARKERS OF SEX DIFFERENTIATION
Edited by Florence P. Haseltine, Michael E. McClure, and Ellen H. Goldberg

REPRODUCTIVE TRACT INFECTIONS: Global Impact and Priorities for Women's Reproductive Health
Edited by Adrienne Germain, King K. Holmes, Peter Piot, and Judith Wasserheit

STEROID CONTRACEPTIVES AND WOMEN'S RESPONSE: Regional Variability in Side-Effects and Pharmacokinetics
Edited by Rachel Snow and Peter Hall

UTERINE AND EMBRYONIC FACTORS IN EARLY PREGNANCY
Edited by Jerome F. Strauss III and C. Richard Lyttle

A Continuation Order Plan is available for this series. A continuation order will bring delivery of each new volume immediately upon publication. Volumes are billed only upon actual shipment. For further information please contact the publisher.

STEROID CONTRACEPTIVES AND WOMEN'S RESPONSE

Regional Variability in Side-Effects and Pharmacokinetics

Edited by

Rachel Snow
Harvard School of Public Health
Boston, Massachusetts

and

Peter Hall
World Health Organization
Geneva, Switzerland

SPRINGER SCIENCE+BUSINESS MEDIA, LLC

Library of Congress Cataloging-in-Publication Data

Steroid contraceptives and women's response : regional variability in
 side-effects and pharmacokinetics / edited by Rachel Snow and Peter
 Hall.
       p.   cm. -- (Reproductive biology)
     "Proceedings based on a Symposium on Steroid Contraceptives and
 Women's Response: Regional Variability in Side-Effects and Steroid
 Pharmacokinetics, held October 21-25, 1990, in Exeter, New
 Hampshire"--T.p. verso.
     Includes bibliographical references and index.
     ISBN 978-1-4613-6039-1   ISBN 978-1-4615-2445-8 (eBook)
     DOI 10.1007/978-1-4615-2445-8
     1. Oral contraceptives--Side effects--Congresses. 2. Oral
 contraceptives--Pharmacokinetics--Congresses. 3. Women--Physiology-
 -Congresses. 4. Steroid hormones--Derivatives--Congresses.
 I. Snow, Rachel. II. Hall, Peter, 1943-   . III. Symposium on
 Steroid Contraceptives and Women's Response: Regional Variability in
 Side-Effects and Steroid Pharmacokinetics (1990 : Exeter, N.H.)
 IV. Series.
     [DNLM: 1. Contraceptive Agents, Female--adverse effects-
 -congresses. 2. Contraceptive Agents, Female--pharmacokinetics-
 -congresses. 3. Steroids--pharmacokinetics--congresses.   QV 177
 S839 1994]
 RG137.5.S756 1994
 615'.766--dc20
 DNLM/DLC
 for Library of Congress   94-21617
                                                            CIP

Proceedings based on a Symposium on Steroid Contraceptives and Women's Response:
Regional Variability in Side-Effects and Steroid Pharmacokinetics,
held October 21–25, 1990, in Exeter, New Hampshire

ISBN 978-1-4613-6039-1

©1994 Springer Science+Business Media New York
Originally published by Plenum Press, New York in 1994
Softcover reprint of the hardcover 1st edition 1994

PREFACE

Pharmacokinetic variability of contraceptive steroids is a relatively under-explored area of contraceptive research, and hardly a common point of discussion among those who plan and deliver family planning services. Nevertheless, numerous independent studies over the last 15 years have indicated that women in different regions of the world vary in their pharmacokinetic response to contraceptive steroids. The causes of such variability are not known, but it has important consequences for contraceptive effectiveness. It may also offer insight to the basis of contraceptive side-effects.

The impetus for this volume was to collect documentation of pharmacokinetic variability of contraceptive steroids, and to explore both the possible causes and implications of these data. Factors known to affect steroid pharmacokinetics, such as concurrent use of specific medications, are reviewed by Back and Orme. Other factors known to affect *endogenous* steroid dynamics are presented in chapters by Bradlow, Longcope, Goldin and Snow, because of their possible role in contraceptive steroid pharmacokinetics.

Regarding the clinical consequences of pharmacokinetic variability, different blood levels of depo medroxyprogesterone acetate (DMPA) and norethisterone enanthate (NET-EN) among Mexican and Thai women have been associated with differences in contraceptive effectiveness; likewise, blood concentrations of levonorgestrel (LNG) among Norplant[R] users are associated with effectiveness of the method. Regrettably, there is little other data regarding the implications of pharmacokinetic variability. For example, we do not know whether variability in blood concentrations of the magnitude observed in these studies are associated with differences in physiologic parameters (such as endometrial affects) or with other side-effects. To date, studies on pharmacokinetic variability have rarely monitored contraceptive side-effects as pharmacodynamic parameters.

This volume also includes several papers addressing the prevalence of contraceptive side-effects, and more general questions regarding clinet satisfaction with steroid contraceptives. We have included such papers in this volume to underscore our interest in seeing greater attention given to possible associatlions between pharmacokinetic parameters and women's experience of contraceptive side-effects.

As the 1994 Cairo World Population Congress approaches, there is increasing clamor for birth control scientists to integrate their research objectives with efforts to promote women's reproductive health and social well-being. In that regard, this book is timely, as it underscores a defined area of research that would explore the physiologic synergies between women's metabolic response to contraception and their response to such technologies. The multi-disciplinary authorship of this volume also models the kind of scientific-programmatic dialogue that is required for doing meaningful research and bringing the fruits of such research to bear on program delivery.

This volume represents a collaborative effort between Harvard University, the Special Programme for Research, Development and Research Training in Human Reproduction (HRP) at the World Health Organization, and the International Planned Parenthood Federation (IPPF), and we are grateful to the support of Mahmoud Fathalla and Halfdan Mahler for this effort. We are also grateful to Sheldon Segal for his advice and support at several critical junctures in the evolution of this volume.

We also wish to thank Vanessa Bingham, Wayne Chueng, Sarah Hemphill, and Jean Joseph for their technical assistance during the editorial process, and most especially Pat Vann and the staff at Plenum Press for their patient and meticulous production of this book.

Finally, we are grateful to acknowledge support by the Ford Foundation for a meeting on Steroid Contraceptives and Women's Response: Regional Variability in Side-Effects and Steroid Pharmacokinetics, which took place in Exeter New Hampshire in October of 1990, and where many of the chapters were originally presented. Our extended thanks to Jose Barzelatto and others at the Ford Foundation for their interest in this project.

Rachel Snow
October 1993

CONTENTS

LESSONS FROM ENDOGENOUS METABOLISM

IMPLICATIONS FOR POLICY AND PROGRAMS

FUTURE RESEARCH DIRECTIONS

WHY PHARMACOKINETICS? AN INTRODUCTION AND OVERVIEW

Rachel C. Snow, and Lucy E. Wilson[1]

In the last two decades, worldwide contraceptive use has risen dramatically, and much of this increase took place in the developing world. From 1960-65 to 1990 the percentage of couples using contraceptives in developing countries rose from 9 to almost 51 percent (United Nations Population Fund 1991). This increase reflects the adoption of three principle types of contraceptives: sterilization, intra-uterine devices, and hormonal methods. Of the approximately 400 million persons contracepting in the late 1980's, 155 million were sterilized, 61 million used hormonal methods and 80 million used the IUD. Different methods appear to be favored (either by providers or clients) in different regions. The global popularity of sterilization and the IUD reflects the popular use of these methods in Asia and the Pacific, the region where 64% of the 400 million contraceptors are living. Yet hormonal methods account for 31% of total use in the Americas and they account for 50% of all method use in Africa (Mauldin and Segal 1988). A recent report on contraceptive use in Sub-Saharan Africa (SSA) suggests that as countries shift from low contraceptive prevalence to higher levels, traditional methods give way to hormonal contraception, followed by a rapid adoption of female sterilization (National Research Council 1993).

Steroid methods of contraception offer numerous features to recommend them, and they are likely to remain a significant offering of family planning services for many years to come. They are highly effective, they incur few, if any, medical risks for healthy women, they are available with different modes of administration, some with a range of dosages, and they generally require low maintenance by the user.

The problem of side-effects

But steroids contraceptives, for all their benefits, are blunt instruments, and many women experience side effects with steroid contraception that are prohibitive, or at best, discouraging. Among contraceptive users in Thailand, reports of side effects are highest among women using injectable steroids (almost 41 percent) and oral contraceptives (38 percent) (Stephen and Chamratrithirong 1988). Among women surveyed in eight developing countries (Grubb 1987), from 26 to 80 percent of users had stopped taking oral

[1] R. Snow, L. Wilson • Department of Population and International Health, Harvard School of Public Health, 665 Huntington Avenue, Boston, MA 02115.

contraceptives because of a fear of side-effects, and a lower or similar percentage of users reported that they had not adopted the pill for such reasons (Grubb 1987). Unfounded fears of health risks associated with contraceptive steroids are not uncommon, and unfounded or not, such perceptions can dramatically affect the acceptability of steroid methods (Grubb, 1987; Mitchell 1994). Even if erroneous fears can be ameliorated through better information, the actual experience of side-effects in the first months of use affects the acceptability of hormonal contraceptives (Basnayake et al 1984). Experience of side-effects such as headache, nausea, vomiting and dizziness in the first two cycles of pill use are strongly predictive of discontinuation (Basnayake et al 1984). In fact, a client's actual experience of side-effects may be more pertinent to a woman's decision to discontinue the method than any abstract notions of potential lethal affects, and the biological and/or social bases of side-effects warrant further investigation.

Regional trends in side-effects

Papers in this volume and elsewhere suggest that there are regional trends in the reporting of contraceptive side-effects. An International Planned Parenthood Federation field trial in six countries (Huezo 1990) indicates that side effects reported for a given method cluster by country. While 66.7 percent of women in a Guatemala clinic reported bleeding problems with injectable contraceptives, only 19.5 percent of women in Kenya, and 29.7 percent of women in Nepal reported such problems with the same method. The percentages of women experiencing amenorrhea, delayed, or scanty bleeding also varied according to geographical location. Regional clustering of gastrointestinal side effects was also noted, with 10.5 percent of Nepalese women reporting such problems following injections, while women from Kenya and Guatemala reported little or no such effects.

E.M. Belsey has reported regional variations in the bleeding patterns among women using hormonal contraceptive methods (1988 and in this volume). Comparing women's bleeding response to a range of different methods (including the vaginal ring, three different types of combined oral contraceptives, two types of progestogen-only pills, and two different doses of DMPA injectables), she and her co-investigators found that within each individual contraceptive method, geographic residence was more closely related to bleeding patterns than any other variable measured. For example, across all methods, Latin American women had short bleeding episodes and long intervals of no bleeding, while European women had more frequent bleeding and spotting days than women from any other region. Use of DMPA was associated with the most varied regional response among women: 72 percent of North African women had amenorrhea on the method, compared to only 25 percent of European women.

Perceptions and reporting of side-effects may reflect cultural preconditions of illness, poor services, low motivation to contracept, or underlying biologic experience. Comprehensive investigations of contraceptive side-effects would ideally weigh all these options, but such in-depth research is not so easy. Regional and cultural differences in the expression of contraceptive experience make it difficult to compare dissatisfaction between populations. Women may incorporate their experience of steroid contraceptive methods with traditional beliefs about fertility, health and the body. Concerns surrounding symptoms may be described in words and expressions that are completely different from the checklist of symptoms provided for the clinical investigators. This is due, in part, to the culturally specific nature of language, but also because a woman's experience of discomfort is conditioned by the impact such discomfort has on her life, and weighed against the advantages provided by the contraceptive.

Furthermore, it is problematic to evaluate women's response to contraceptive methods through reported dissatisfaction (or certainly discontinuation), when the quality of services is so variable between regions, and even within the same region (see chapters by Kaufman and Bruce in this volume). Poor service quality confounds comparisons of women's experience between regions, because dissatisfaction may be a function of what alternative methods and services are available. How does one account for differences in women's fortitude for tolerating a specific method, if that is the only method available?

A premise of this volume is that greater attention should be paid to evaluating the plausibility that side-effects clustered by region actually reflect something biological. Endogenous steroid dynamics (Tyrer et al 1973), body weight (Talwar and Berger 1977) and nutrition (Measham et al 1980) have been each been associated with the manifestation of contraceptive side-effects, and each of these factors vary by region and population.

Furthermore, a slim body of literature has accumulated over the last 15 years illustrating regional variability in the pharmacokinetics profiles of contraceptive steroids. In this respect, contraceptive steroids are not unlike many other drugs, for which variable metabolic responses have been noted in different client populations (Hvidberg 1990). The aim of this volume is to provide the interested reader with an overview of the variability in pharmacokinetics profiles that are observed for contraceptive steroids, and to weigh possible causes and consequences of such variability for women's experience of these family planning technologies.

Pharmacokinetic variability by region

Pharmacokinetic response can only be detected serologically, and therefore pharmacokinetics is most often a silent dimension of women's response to steroid contraceptives. Pharmacokinetic studies are often small, with 10 or fewer women observed in each client population. Given such small numbers, observations of distinct regional patterns are all the more striking.

Plasma levels of ethinyl estradiol (the principal estrogen used in oral contraceptives) were evaluated in several populations following administration of a single dose, and variations among the population groups were reported by Goldzieher and colleagues (1980). The highest plasma concentrations were observed among Thai women, and the lowest among Nigerian women; the imputed area under the serum concentration curve (AUC) for Thai women was higher than those predicted for lower doses of the steroid among women in the USA; these data were corrected for differences in body surface (Goldzieher et al 1980).

Looking at data on the two-month injectable norethisterone enanthate (NET-EN), plasma concentrations of NET following a single injection were detectable for a much shorter time among women in Delhi, India than they were among women in Sweden

(Fotherby et al 1980); the apparent elimination time following intramuscular injections of NET-EN was also found to be significantly shorter among British, relative to Chinese women (Sang, personal communication).

Garza Flores and colleagues (see chapter in this volume) have shown regional differences in the pharmacokinetics of intramuscularly injected DMPA. In a comparison of the DMPA pharmacokinetics in Thai and Mexican women, Thai women displayed a greater area under the serum concentration curve (AUC) and a shorter elimination half life ($t_{1/2}$). In addition, Thai women showed a faster run to ovulation than Mexican women, as well as a greater sensitivity to the particle size of steroid preparation. In a seperate study by the same investigators, Mexican women receiving intramuscular injections of NET-EN showed a higher C_{max} than women in the USA, UK, Sweden, and India.

Causes and consequences of pharmacokinetic variability

There is evidence to suggest that some component or correlate of body weight may be associated with steroid pharmacokinetics. Support for such an association comes from some direct measures of blood concentrations and body weight, but also from indirect data on weight and effectiveness.

In initial studies by Garza-Flores and colleagues, thin women in Thailand and Mexico appeared to absorb intramuscular injected DMPA more rapidly than obese women, but regional differences in pharmacokinetic profiles between Thailand and Mexico eclipsed any detectable differences by body mass index (Garza-Flores et al in this volume).

Blood concentrations in the fifth year of Norplant[R] use vary by weight of the user; and the gross pregnancy rates at the end of five years of Norplant[R] use range from 0.2 in women less than 50 kg to 8.5 in women greater than 70 kg (Sivin 1988; Nash in this volume). In contrast, the body weights of Indian women who became pregnant on NET-EN were, on average, 7 kgs *lighter* than women with no method failure (Banerjee et al 1984).

Body weight may be associated with differences in blood concentrations of contraceptive steroids or contraceptive effectiveness because of simple dilution effects, or sequestering of the steroid in fat depots at the site of injection or implant. But Goldzieher's evidence of significant regional differences in the pharmacokinetics of *orally* administered steroids (Goldzeiher et al 1980), with the data normalized for body size, reduce the likelihood that observed differences are due to dilution or sequestering alone.

Under-explored is the possibility that baseline or induced differences in sex hormone-binding globulin (SHBG) may be responsible for regional variability, especially given the variations in SHBG with body weight. SHBG levels are low in obese women and higher in thin women, and SHBG also varies with the shape of fatness. Obese women with a predominance of fat above the umbilicus (android obesity) have lower levels of SHBG than obese individuals with an excess of hip and femoral fat (Evans et al 1983). Fat topography, or the waist to hip distribution of body fat, is variable by ethnicity, and appears to be at least partially inherited (Bjorntorp 1988).

SHBG binds steroids more tightly than albumin, leaving less free non-protein bound, biologically active steroid. While free ethynyl estradiol (EE) and sulphated EE readily bind to albumin, neither bind to SHBG. Conversely, gestagens such as levonorgestrel (LNG) bind to both albumin and SHBG, and gestagens inhibit SHBG synthesis. Fotherby (1988) has illustrated the magnitude of the effect which SHBG can have on serum gestagen concentrations by measuring serum gestagen concentrations in women who have been administered LNG alone versus those administered combined LNG and EE. Those with LNG and EE show serum LNG levels 4 to 5 times higher than those given LNG alone. EE may stimulate SHBG synthesis, to which LNG may then increasingly bind, thus maintaining higher serum LNG levels. While more research is need to explicate the net effect of exogenous steroids on the bound vs free fraction of endogenous steroids, and vice-versa, the available data would suggest that the underlying dynamics between SHBG and the endogenous steroid concentrations may have important effects in determining the pharmacokinetic profile of the contraceptive steroids (Fotherby 1988).

The Importance of Underlying Biologic Conditions

Underlying characteristics of the individual are no doubt important to a woman's ultimate response to exogenous steroids (Harlow in this volume), but more investigations are needed to disaggregate the interactions between endogenous and exogenous systems. Improved understanding of factors responsible for underlying heterogeneity in reproductive function among women may be of use to future studies of contraceptive variability. Indeed, populations factors associated with follicular development and menstruation, such as body composition and diet (see Snow and Goldin in this volume), may warrant greater attention in contraceptive studies even now.

The Implications for Contraceptive Side-Effects

Ultimately, the principal question for the contraceptive research community is whether the observed variability in contraceptive steroid pharmacokinetics has any significance for women's experience of these methods: are varying blood levels of contraceptive hormones, and different absorption or elimination profiles, correlated with users's experience of side-effects?

Several papers in this volume attest to the importance of pharmacokinetic variability for the effectiveness of these methods for select populations (see Garza-Flores, Nash); but there are very few, if any, data wherein pharmacokinetic profiles and side-effects were documented in the same clients. This is partly due to the smallness of the typical pharmacokinetic study, but Sang Guo Wei argues that studies would not be compromised by collecting fewer blood samples on each woman, allowing more subjects to be observed overall (Sang, in this volume).

The data at hand indicate some regional patterns in women's reported experience of contraceptive side-effects, and in the pharmacokinetic profiles of contraceptive steroids. But the data exist in disparate studies, rarely if ever co-joined in the same women. Clearly, more and better data are needed to explicate the interactions between underlying population biology, contraceptive pharmacokinetics, and reported side-effects; in order to explore regional effects, studies will ideally be local and comparative.

The biological bases of user dissatisfaction with steroid contraceptives remains relatively under-explored. As steroid contraceptives are, and will be, increasingly distributed in populations of variable ethnic, geographic and cultural origin, it becomes increasingly important to investigate the extent to which women's experience of side-effects reflect biologic factors that would allow more sophisticated prescriptive regimens and greater client satisfaction.

References

Banerjee SK, *et al.* Comparative evaluation of contraceptive efficacy of norethisterone oenanthate (200mg) injectable contraceptive given every two or three months. *Contraception* 30(6):561-573. 1984.

Basnayake S, *et al.* Early symptoms and discontinuation among users of oral contraceptives in Sri Lanka. *Studies in Family Planning* 15(6):285-290. 1984.

Belsey EM, Peregoudov S. Determinants of menstrual bleeding patterns among women using natural and hormonal methods of contraception. I. Regional variations. *Contraception* 38:227. 1988.

Belsey EM, d'Arcangues C, Carlson N. Determinants of menstrual bleeding patterns among women using natural and hormonal methods of contraception. II. The influence of individual characteristics. *Contraception* 38:243. 1988.

Benagiano G, *et al.* Return of ovarian function and endometrial morphology in women treated with norethisterone oenanthate: A pilot study. *Fertility and Sterility* 34(5):456-60. 1980.

Bjorntorp P. Fat cell distribution and metabolism. *Annals of the New York Academy of Sciences* 66-72. 1988.

Commission on Health Research for Development. *Health Research: Essential Link to Equity in Development.* New York: Oxford University Press. 1990.

Evans DJ, Hoffman R, et al. Relationship of androgenic activity to body fat topography, fat cell morphology, and metabolic aberrations in premenopausal women. *J Clin Endocrinol Metab 57:304-10. 1983.*

Fotherby K. Interaction of contraceptive steroids with binding proteins and the clinical implications. *Annals of the New York Academy of Sciences* 538:313-324. 1988.

Fotherby K, *et al.* A preliminary pharmacokinetic and pharmacodynamic evaluation of depot-medroxyprogesterone acetate and norethisterone oenanthate. *Fertility and Sterility* 34(2):131-139. 1980.

Goldzieher J. Plasma levels and pharmacokinetics of ethinyl estrogens in various populations. *Contraception* 21(1):1-15. 1980.

_____. Selected aspects of the pharmacokinetics and metabolism of ethinyl estrogens and their clinical implications. *Pharmacodynamics and Pharmacokinetics* 163(1):318-322. 1990.

Grubb G, co-ordinator. Women's perceptions of the safety of the Pill: A survey in eight developing countries. *Journal of Biosocial Science* 19:313-21. 1987.

Herzberg BN, Johnson AL, Brown S. Depressive symptoms and oral contraceptives. *British Medical Journal* 4:142-45. 1970.

Huezo C. Factors affecting method selection and continuation: Preliminary data from a six-country study. Paper presented at Steroid Contraceptives and Women's Response:

Regional Variability in Side-Effects and Steroid Pharmacokinetics, Exeter, New Hampshire, October 21-24, 1990.

Hvidberg E. Why do we need pharmocokinetic studies? *Pharmacodynamics and Pharmacokinetics* 163(1):316-318. 1990.

Kane FJ. Evaluation of emotional reactions to oral contraceptive use. *Am J Obstet Gynecol* 126(8):968-72. 1976.

Lan PT. Return of ovulation following a single injection of depo-medroxy-progesterone acetate: A pharmacokinetic and pharmacodynamic study. *Contraception* 29(1):1-18. 1984.

Mauldin WP, Segal S. Prevalence and contraceptive use: Trends and issues. *Studies in Family Planning* 19(6):335-353. 1988.

Measham AR, Khan AR, Huber DH. Dizziness associated with discontinuation of oral contraceptives in Bangladesh. *International Journal of Obstetrics and Gynecology* 18(2):109. 1980.

National Research Council. *Factors Affecting Continued Use in Sub-Saharan Africa.* Washington D.C.: National Academy Press. 1993.

Schwartz JB, Flieger W. Contraceptive prevalence and continuation: A longitudinal analysis of traditional and other method users in the Philippines. In: *Dynamics of Contraceptive Use.* Tsui AO, Herbertson MA, eds. Cambridge, England: Parkes Foundation. 1989.

Sivin I. International experience with Norplant and Norplant-2 contraceptives. *Studies in Family Planning* 19(2):81-94. 1988.

Snow R, Schneider BA, Barbieri R. High dietary fiber and low saturated fat intake among oligomenorrheic undergraduates. *Fertility and Sterility* 54(4):632-637. 1990.

Stephen EH, Chamratrithirong A. Contraceptive side effects among current users in Thailand. *International Family Planning Perspectives* 14(1):9-14. 1988.

Talwar PP, Berger GS. The relation of body weight to side effects associated with oral contraceptives. *British Medical Journal* 1(6077). 1977.

Tyrer LB, Isenman AW, Knox EG. eds, *Seminar on Family Planning.* Chicago: American College of Obstetricians and Gynecologists. 1973

United Nations Population Fund. New York. 1991.

Wyshak G, Snow R. Fiber consumption and menstrual regularity in young women. *The Journal of Women's Health* 2(3):295-299. 1993.

SELECTION OF A CONTRACEPTIVE: WHAT GUIDES A WOMAN?

Pramilla Senanayake [1]

1. INTRODUCTION

The modern contraception revolution is now some 30 years old. From its beginning the provision of contraceptives has been closely guarded by the health care professionals. There were, and are, many reasons for this. However, this paper will not carry out an analysis of these factors.

Most health care professionals, especially in the 60s and 70s, had training in, and concentrated on, the area of curative medicine. They were accustomed to making decisions on behalf of their patients. In most cases, these decisions were subsequently accepted by the patients as "the patients' decisions." For example, a doctor who examines a patient and diagnoses acute appendicitis may then go on to suggest an appendectomy. The patient, who by now is in acute pain, feels that he has little choice in the matter, and indeed may feel that he does not need a choice as, under the circumstances, the doctor knows best. He goes on to accept the doctor's offer of an appendectomy, in terms of "I chose to have my appendix removed."

In the area of preventive health care the scenario should be quite different. Here we are not dealing with sick people, we are dealing, in the majority of cases, with normal, healthy men and women who may wish to follow a particular course of action (request immunization against a certain disease, use contraception, etc.) which, in general, will better their state of health or improve their well-being. In spite of this, and even today, in many family planning clinics and large-scale family planning programmes, what guides a woman

P. SENANAYAKE • International Planned Parenthood Federation, Regent's College, Inner Circle, Regent's Park, London NW1 4NS, United Kingdom

Steroid Contraceptives and Women's Response,
Edited by R. Snow and P. Hall, Plenum Press, New York, 1994

into selecting a contraceptive are factors over which she may not have any direct control. For example, a woman who selects an injectable contraceptive may never be able to use it, because though it is available in certain countries, she may not be fortunate enough to live in a country where the method is approved as a contraceptive and where it is freely available.

2.   WHAT THEN GUIDES A WOMAN IN THE SELECTION OF HER CONTRACEPTIVE?

This paper deals at some length with the role that the medical/health care professional plays in guiding that decision. It concludes by considering some of the factors that should in reality be the guiding factors in the decision-making process for the individual and the couple.

Much has been written about how important it is that the eventual method of contraception chosen is acceptable to the individual and to the couple. Also, it is pointed out that all methods should be clearly explained so that each couple is in a position to choose the method that best suits them. Nevertheless, it is apparent that health workers spend a great deal of time constructing ideal types of individuals and matching them with appropriate methods. What is considered ideal is not only based on physiological and medical factors, but also on social and cultural factors.

The historical development of the distinction between medical and social factors is interesting. In the early stages of the contraceptive revolution contraceptive services centered around a more physiological interpretation of the proper concern of medicine. However, more recent debates reflect the growing concern of medicine as a whole with the social aspects of medicine. What needs to be considered is the medical profession's knowledge of social factors. The health workers' perceptions of the individual's contraceptive needs are based on assumptions about the social organization surrounding reproduction which supports existing social and cultural norms. Medicine, in turn, reinforces these norms in its construction of "patient choice." Little attention is paid to the needs of the individuals who might end up having difficult reproductive histories.

What is the medical view of the function of contraception, and what are the factors which constitute the "medical construction" of the contraceptive needs of their "patients?" Decisions made by doctors are influenced not only by the state of medical knowledge, but also by their perceptions of social and cultural norms. What is the function of contraception and what is it seen to be preventing? What are the criteria by which methods are chosen for the "patient?" What are the stages in reproductive life that the medical profession assumes are associated with changing contraceptive needs?

The functions of contraception have been identified as preventing pregnancy, preventing the dangers to health from pregnancy or fears of pregnancy, avoiding having births spaced too closely together, preventing grandmultiparity, preventing illegitimacy, preventing abortion, preventing over-population, and reducing the costs to the community of unwanted pregnancies.

In the light of the above functions, certain other criteria, such as physiological factors, need to prevent pregnancy, motivation of the client, and his/her sexual behavior and sexual relationships, are taken into consideration.

2.1 Physiological Factors

Physiological criteria that will play a part in the decision-making process of the health worker will include: for example, age of the individual, if an adolescent, or an older woman say in her 40s; whether or not there is a need to maintain future fertility; and whether or not a woman is breast feeding.

2.2 Need to Prevent Pregnancy

The health worker may assume that those who are using contraception for spacing between pregnancies do not necessarily need the most effective contraceptive methods. This however may be far from the truth. For these same individuals spacing may be very important for reasons such as an inability to care for the older child, need to resume a career, and/or the inability to obtain or accept abortion. Often in the eyes of the medical profession the need to avoid pregnancies seem to be related to the marital and reproductive status of the individual rather than to his/her aspirations.

2.3 Motivation

In the literature and in research studies, there is a great deal of emphasis placed on motivation to use contraception and the motivation to continue to use particular methods. In the literature motivation is a blanket term, a word used to describe whether or not an individual seems likely to use the contraceptive method in question. It is not an explanatory term in that it does not explain why individuals do or do not use contraception at various times. For example, consider the following comment: "The loop or coil is the method of choice for the unmotivated patient, giving the maximum protection with the minimum effort."

The term "feckless" was often used to describe these women. "Feckless females" is a generic term used to cover the type of women all GPs know well. Of inadequate personality, low intelligence, and prone to anxiety, they are bound to miss pills or stop taking them without seeking medical advice. Side effects can be guaranteed to be severe and multitudinous. Recurrent unwanted pregnancies and pitiful requests for abortion will result.

Even though the above comment is from literature published some 15 years ago, it is unfortunate that in this respect the situation has not changed dramatically. The "user perspective" has been a fashionable term from the late 1970s onward. Even though much has been written about the importance of considering the views, needs and perspectives of the individual who contemplates the use of contraception, many family planning programs pay only lip service to this aspect.

2.4 Sexual Relationships

Literature abounds with references to the implications of the different types of sexual relationship for the choice of contraceptive method. However, there is a little or no reference to the effect of different types of contraceptive methods on the developing relationship. At most, the discussion on sexual behavior revolves round frequency of sexual intercourse, spontaneity and aesthetics. In discussing the implications of what the medical profession perceives as different sexual relationships, perhaps requiring different methods of contraception, the pre-existence of a relationship is assumed. What most practitioners fail to take into account is that using or discussing contraception once a relationship is established is very different from using or preparing to use methods in the very early stages of a relationship, i.e., before a relationship is established and before first coitus.

Assuming the existence of a heterosexual relationship, the literature does make a distinction between "different stages" in a relationship. This becomes quite evident in the discussion of coitus-independent and coitus-dependent methods. The Pill is said to have particular value where there has been no previous sexual experience. Oral contraceptives provide the couple with full spontaneous sexual expression at all times. On the other hand, once a couple have established a satisfactory sexual relationship they are more likely to tolerate the lack of spontaneity inevitable when a diaphragm or condom is used.

Elsewhere in the literature it is stated that well-adjusted women are able to use the diaphragm in such a way that the natural spontaneity of the relationship is not spoilt.

Coital frequency is a subject which occurs in the discussion of condom use. There is general agreement that for women who do not have intercourse frequently, methods such as the pill and the IUD are inappropriate. However, in reality this may be far from the truth. In relationships where one partner is away for long periods of time, the couple may place a high premium on spontaneity and high coital frequency when they are together. There is also little thought given to the reality of the life of the woman who has no single established relationship, and who has sporadic intercourse. She still needs protection and is not able to predict when she may need this. She needs infrequent protection, and yet perhaps needs a method which she can use herself and which is not coitus-dependent thus obviating the need to discuss contraception with a potential partner.

What then are the attributes of a method of contraception that guide an individual in selecting a particular method? In a study carried out by Poo-King Kee and Russel K.

Darroch the authors used a psychological measurement technique as a measure of perception of contraceptive methods. Bipolar adjectives were used. These covered such attributes as **effectiveness** (effective/ineffective, efficient/inefficient, reliable/unreliable, successful/unsuccessful, safe/unsafe); **discretion** (discreet/obvious, invisible/visible, obtrusive/unobtrusive, messy/non-messy, quick/time-consuming); **pleasure** (exciting/boring, embarrassing/non-embarrassing, distressing/non-distressing, attractive/ugly, satisfying/frustrating, comfortable/uncomfortable, painful/painless, stress-free/stressful; **legality** (legal/illegal, moral/immoral); **permanence**; **cost** (expensive/inexpensive); access and availability (easy/difficult).

The acceptability or otherwise of a contraceptive may thus be summarized on certain attributes of the method itself. There are the relative importance of the attributes in perceptions between males and females. Success or failure in the acceptance of contraception depends to a large extent on the identification of these attributes and their subsequent communication to the administrators of the family planning programs.

In addition to the method-related attributes, the nature, quality and efficiency of the services are probably of equal importance in determining the acceptability and continued use of a method. The choice of a particular method is strongly influenced by the methods available through the family planning program, or promoted through the use of target systems, but it is not dependent on these aspects alone.

The attributes of the service providers, including their attitudes; personal experiences of, and preferences for, particular contraceptives; quality, length and method of counselling; the regime of clinical examination and follow-up visits; the treatment of side effects; and the scale of fees charged may all have a direct bearing on the type of contraceptive selected.

The attitudes and the beliefs of the spouse may be as important a determinant of adoption and continuation as those of the user. The willingness of the husband and wife to talk about sex and contraception may also be an important influence on successful use, particularly of coitus-related methods.

Among the various determinants of contraceptive choice, general beliefs relating to health may also play an important part. Beliefs about the function of menstruation, and erroneous beliefs about the female anatomy may give rise to fears that IUDs or barrier methods can get "lost" in the body (MacCormac 1985). Attitudes toward touching and cleaning the genital areas may also carry implications for the convenience and use-effectiveness of certain methods.

3. CONCLUSIONS

In the 1990s the number of fertile women in the third world countries will increase by one third, and if, at the same time, contraceptive prevalence is to continue rising, then the number of couples using contraception needs to double between now and the year 2000.

Service providers can assist in achieving this objective if they pay careful attention to clients' wishes with regard to selection of a contraceptive.

We do not, as yet, have an "ideal" or "perfect" contraceptive. We may never have a method that suits all individuals all the time. However, we do have a range of effective and efficient methods. If these methods are made more widely available, their use, mode of action, advantages and disadvantages are clearly explained, and the service providers are willing to listen to the clients and provide them with methods and services that the clients select, then family planning programs will be in a better position to meet the challenges of the 1990s.

CONTRACEPTIVES AND WOMEN'S COMPLAINTS - PRELIMINARY RESULTS FROM THE POST-MARKETING SURVEILLANCE OF NORPLANT[R]

International Collaborative Surveillance of Norplant[R]

Y. Ahmed, S. Boccard, T. Farley, O. Meirik[1]

1. INTRODUCTION

Since 1975, when clinical trials on the contraceptive device Norplant[R] were initiated by the International Committee for Contraceptive Research (ICCR), a large number of studies of Norplant[R] have been conducted to address issues varying from use/effectiveness and acceptability of Norplant[R] among different populations to studies on release rates of levonorgestrel from the device, its specific effects on various organ functions, side effects and continuation rates. These studies were undertaken in different populations and at different times. Accumulated experience from these studies has been reviewed by WHO (WHO 1985) and Sivin (Sivin 1988).

Higher pregnancy rates have been reported in well-nourished users of Norplant[R] populations from developed countries as compared to women living under less privileged circumstances in developing countries. The underlying reason for the difference in pregnancy rates is thought to be due to differences in body weight of the studied populations (Sivin 1988). It is also reported that bleeding disturbances with Norplant[R] may be related to body weight (Sivin I. Unpublished 1990). Conceivably, differences between populations and ethnic groups with regard to the occurrence of side effects and of efficacy

O. MEIRIK • Special Programme of Research, Development and Research Training in Human Reproduction, World Health Organization, 1211 Geneva 27, Switzerland

of a hormonal contraceptive such as Norplant[R] could be mediated through different metabolic characteristics influenced by environmental, dietary or genetic factors. For example, in the mid-1960s when oral contraceptives (OCs) were approved in many countries in Europe, there were several conflicting reports on pathological liver function tests among women using OCs (Swyer & Little 1965, Larsson-Cohn 1965, Eisalo et al 1965). The reason for the varying results of the studies was never identified but it is interesting to note that the studies where abnormal results of liver tests were found were undertaken in the Nordic countries while studies from other parts of Europe generally showed no change, suggesting that environmental or genetic factors or both may have been implicated. Differences in various aspects of health exist between different ethnic groups (Cruikshank & Beevers 1989). Differences between ethnic groups in culture, education, diet and personal habits may be reflected in different manifestations of diseases, different disease patterns, or different perceptions of normal body functions.

The observation of different failure rates of Norplant[R] across populations and a correlation between bleeding disturbances and body weight for users of Norplant[R] prompted us to evaluate whether the type and rate of complaints of users of Norplant[R] vary across populations. For this preliminary report we used data on various complaints according to contraceptive method as reported by women in four countries participating in an international multicentred study on the "Post-marketing Surveillance of Norplant[R]." This post-marketing surveillance is currently being carried out in eight developing countries where Norplant[R] is or is anticipated to be approved. The project is designed to identify short to medium term side effects of Norplant[R] which have not been identified in clinical trials. This paper is based on the preliminary one-year data that was available for analysis from four of the eight countries participating in the project. Our intention was to study differences of rates and types of complaints reported by women in various countries and to assess if such complaints might serve as indicators for differences in pharmacological effects of the levonorgestrel, the effective agent of Norplant[R].

2. METHODS AND MATERIALS

Data for this report are from the "Post-Marketing Surveillance of Norplant[R]," a prospective cohort study aimed to identify possible short to medium-term side effects of Norplant[R] that had not been seen or noticed in clinical trials. Women who attended family planning clinics and chose Norplant[R], IUD or surgical sterilization were invited to the study. They were enrolled if they accepted and fulfilled the screening criteria which corresponded to the contraindications for use of Norplant[R] and IUD. At admission the women were interviewed using a structured questionnaire. Data recorded from the interview comprised personal history, including education, occupation, marital status, smoking and use of alcohol; and reproductive, contraceptive and medical history. A physical examination was done including a gynecological examination, cervical smear, hemoglobin and blood pressure. For each index subject (Norplant^R acceptor) one control subject (an IUD acceptor or a sterilization acceptor) was enrolled. Each clinic used the type of IUD or sterilization procedure it routinely provided. At each participating clinic the age distribution of the control subjects was kept similar to that of index subjects by means of frequency matching by age in five-year categories. Apart from the usual visit one to four weeks after insertion of Norplant^R or the IUD, or surgery for sterilization, the women were asked to return to the clinic at 6 monthly visits for structured interviews and physical examination if clinically indicated. They were also invited to return to their family planning clinic for any health problems or complaints they may have experienced between the regular six-month scheduled visits. Such unscheduled visits, the reasons for them, and the diagnoses, were recorded in the context of the study.

Symptoms and complaints reported by the women were coded according to the International Classification of Primary Care (ICPC) (Lamberts & Wood 1988). ICPC includes a component relating to reasons for the encounter with the health care system. This component of ICPC contains a number of chapters containing descriptions and coding for general unspecific symptoms and complaints, and complaints and symptoms relating to organs and organ systems, e.g., "blood, blood-forming organs, lymphatics, spleen" or "ear," and to functional systems such as "endocrine, metabolic and nutritional" or "neurological." It also includes complaints relating to "social problems," and "psychological," and to requests for services, as in the chapter "Pregnancy, Childbearing, Family Planning." The recording of symptoms and complaints was done by the social worker, nurse/midwife or physician who interviewed the woman at the routine ordinary follow-up or unscheduled visits. As a rule the interviewer discussed the complaint and symptom and asked the woman if she agreed to the classification before it was recorded.

End points in the analysis were overall complaints and symptoms, and complaints and symptoms related to the contraceptive methods under study according to known side effects of the Norplant^R and IUDs. Complaints that were related to the methods were for example bleeding and menstrual disturbances, low abdominal pain and genital complaints in general. Depression, anxiety, headache and vertigo were also classified as method-related complaints since these complaints have been reported to be associated with the use of Norplant^R in earlier studies (WHO 1985, Sivin 1988).

In this preliminary analysis we used simple cross-tabulations by country, contraceptive method, and weight. The ratio of the rate of complaints by Norplant^R users and IUD users was computed in some instances to adjust for the assumed effect of clinic and country on the propensity to make complaints and to record them.

For the purpose of this analysis women who continued their initial contraceptive method through one year of follow-up were included. Although by July 1990 some 12,000 women had been enrolled for follow-up in seven countries, the number of women for

Table 1. Number of women according to contraceptive method and country.

Country	Contraceptive method			
	Norplant[R]	IUD	Sterilization	All
Chile	246	198	15	444
China	227	194	-	421
Sri Lanka	200	82	100	282
Thailand	382	333	10	725
All	1.055	807	125	1.987

Table 2. Mean and 1 standard deviation in parenthesis, of age, parity, age at menarche, weight (kg) and height (cm) of women by contraceptive method and country.

	Country	Contraceptive method		
		Norplant[R]	IUD	Sterilization
Age	Chile	27.5 (4.5)	28.3 (4.8)	32.8 (2.9)
	China	30.0 (3.5)	30.1 (3.8)	
	Sri Lanka	25.5 (5.2)	27.4 (4.7)	28.5 (4.6)
	Thailand	26.7 (4.5)	26.6 (4.2)	31.0 (4.3)
	All	27.4 (4.7)	27.9 (4.5)	29.2 (4.7)
Parity	Chile	2.1 (1.0)	1.9 (0.8)	4.1 (1.0)
	China	1.0 (0.2)	1.0 (0.1)	
	Sri Lanka	2.0 (1.0)	1.8 (0.9)	3.2 (0.9)
	Thailand	1.6 (0.7)	1.3 (0.6)	2.4 (0.7)
Age at	Chile	13.1 (1.6)	12.9 (1.6)	12.7 (1.2)
menarche	China	14.3 (1.6)	14.3 (1.7)	
	Sri Lanka	13.7 (1.5)	13.3 (1.6)	13.5 (1.2)
	Thailand	14.3 (1.5)	14.0 (1.4)	14.7 (1.7)
Weight	Chile	57.0 (8.3)	60.3 (8.6)	60.9 (10.2)
Kg	China	55.6 (6.9)	56.6 (7.5)	
	Sri Lanka	40.5 (6.7)	45.8 (8.8)	43.9 (7.7)
	Thailand	53.3 (8.1)	53.8 (9.1)	48.7 (4.8)
Height	Chile	156.1 (5.7)	156 (5.7)	154.7 (2.5)
	China	160.9 (5.1)	160.6 (4.4)	
	Sri Lanka	150.5 (6.0)	151.9 (5.6)	151.5 (5.6)
	Tahiland	153.7 (4.9)	154.2 (5.3)	154.0 (2.5)

whom checked and computerized data for one year of follow-up were available was 1,987. The women were from Chile, Peoples Republic of China, Sri Lanka and Thailand. Of these, 1,055 were Norplant^R users, 807 used IUDs and 125 had been sterilized. **Table 1** shows the distribution of women according to country and contraceptive method. Since the number of acceptors of sterilization was small, we restricted this preliminary analysis to Norplant^R and IUD acceptors only.

3. RESULTS

Means and standard deviations of age, parity, age at menarche, and weight and height at admission for the women by country and contraceptive method are shown in **Table 2**. There were no important differences among women within countries except that Norplant^R acceptors in Chile and Sri Lanka were lighter than the normal subjects. Women in Sri Lanka were between 10 and 15 kg lighter than women in other countries. **Table 3** shows the occurrence of overall complaints. The rate of complaints is higher for index subjects compared to controls in all four countries. In Sri Lanka as many as 97 percent of Norplant^R acceptors had at least one complaint recorded during the first year of use. A very large proportion of the complaints made could be classified as method-related complaints, these are shown in **Table 4**.

Table 5 shows the number of women with specified complaints made during the first year and rates thereof. The types of complaints are given in the heading of the table. The rates of complaints show considerable difference between Norplant^R and IUD users. The rates differ also among countries as was also seen in **Tables 3 and 4**. However, the rate-ratio (ratio of rates of complaints for Norplant^R acceptors and IUD acceptors) show far fewer differences among the countries. This indicates that a large proportion of the difference between countries could be attributable to effects of country and clinics. Worth noting is the similarity of the rate ratios of irregular and intermenstrual bleeding complaints among the four countries (Complaint D), and the differences in rate ratios of excessive menses (complaint C) (P<0.01). There are high rates of complaints of excessive menses among IUD users in China and similarly high rates of this complaint in Sri Lanka for both IUD and Norplant^R users.

Table 6 shows the same distribution of complaints as in **Table 4** but for complaints made in the second half of year of use of the methods. Typically, the rates of complaints for all types are lower when the second half of the year is considered. The same pattern emerges from this table as in the previous. Worth noting are the high rates of bleeding disturbances among IUD users in the Chinese centres and the significant heterogeneity test for complaints of excessive menses (Complaint C).

Tables 7A and 7B show the number of women with various method-related complaints, and crude and weighted rate-ratios, by weight of the women. The rate- ratios

Table 3. Numbers and rates of women with one or more complaint during 1st year of use of Norplant^R & IUD by method & country & ratio of complaint rate of Norplant^R & IUD users.

Country	Contraceptive method						Rate-ratio
	Norplant^R			IUD			
	All women	With complaints		All women	With complaints		
Chile	246	182	74.0%	198	108	54.5%	1.4
China	227	166	73.1%	194	102	52.6%	1.4
Sri Lanka	200	194	97.0%	82	47	57.3%	1.7
Thailand	382	141	36.9%	333	98	29.4%	1.3
All	1.055	683	64.7%	807	355	44.0%	1.5
Adjusted rate-ratio							1.4
confidence limits							1.2-1.6

Table 4. Numbers and rates of women with one or more method related complaint during 1st year of use of Norplant^R and IUD by contraceptive method and country, and ratio of complaint rate of Norplant^R and IUD.

Country	Contraceptive method						Rate-ratio
	Norplant			IUD			
	All women	With complaints		All women	With complaints		
Chile	246	163	66.3%	198	90	45.5%	1.5
China	227	151	66.5%	194	72	37.1%	1.8
Sri Lanka	200	192	96.0%	82	44	53.7%	1.8
Thailand	382	134	35.1%	333	83	24.9%	1.4
All	1.055	640	60.7	807	289	35.8	1.7
Adjusted rate-ratio							1.6
confidence limits							1.4-1.8

refer to complaints given during the first year of use. The number of observations are few in some of their cells and in particular for women 70 kilograms (kg) or more. Only 96 women, or fewer than 5 percent of all women weighed 70 kg or more. In the weight category 50 - 90 kg, where most of the women are found, the heterogeneity test is significant for complaints of excessive menses. Judging from the rate-ratios by weight and country (**Table 7C**) there is no consistent association between weight and complaints. In Chile, Sri Lanka and Thailand the rate-ratios of irregular menses and intermenstrual bleeding (Complaint C) appear to be lower among women in the lighter weight categories compared to the heavier, but this is reversed for Chinese women.

Table 5. Numbers, rates and rate-ratios of women with various complaints during 1st year of use, by contraceptive method and country. Complaints are A: headache, weakness, nausea, vertigo, anxiety, irritability, insomnia and loss of libido; B: Scanty or postponed menses; C: excessive menses; and D: irregular menses and intermenstrual bleeding.

Complaint	Country	Contraceptive method NORPLANT			IUD			Rate-ratio
		All women	With complaints	%	All women	With complaints	%	
A.	Chile	246	51	20.7	198	11	5.6	3.7
	China	227	31	13.7	194	6	3.1	4.4
	Sri Lanka	200	17	8.5	82	3	3.7	2.3
	Thailand	382	19	5.0	333	6	1.8	2.8
	All	1,055	118	11.2	807	26	3.2	3.5
	Adjusted rate-ratio							3.4
	confidence limits							2.2-5.3
	x^2							1.0
B.	Chile		67	27.2		14	7.1	3.9
	China		20	8.8		2	1.0	8.5
	Sri Lanka		124	62.0		13	15.9	3.9
	Thailand		84	22.0		37	11.1	2.0
	All		295	28.0		66	8.2	3.4
	Adjusted rate-ratio							2.8
	confidence limits							2.2-3.7
	x^2							7.8[*]
C.	Chile		15	6.1		17	8.6	0.7
	China		8	3.5		26	13.4	0.3
	Sri Lanka		46	23.0		14	17.1	1.3
	Thailand		12	3.1		3	0.9	3.5
	All		81	7.7		60	7.4	1.0
	Adjusted rate-ratio							0.8
	confidence limits							0.6-1.2
	x^2							15.7[**]
D.	Chile		62	25.5		15	7.6	3.3
	China		125	55.1		36	18.6	3.0
	Sri Lanka		115	57.5		15	18.3	3.1
	Thailand		57	14.9		14	4.2	3.5
	All		359	34.0		80	9.9	3.4
	Adjusted rate-ratio							3.2
	confidence limits							2.5-4.0
	x^2							0.3

4. DISCUSSION

Compared to other studies on side effects of Norplant[R] and IUDs, we found high rates of the complaints made by the women studied (ICCR 1978, Sivin 1983, Du et al 1990). Practically all of the Norplant[R] acceptors in Sri Lanka complained about health problems, while only about one third in Thailand did so. The methodology for recording

Table 6. Numbers, rates and rate ratios of women with various complaints during 6-12 months of use, by contraceptive method and country. Complaints are: A: headache, weakness, nausea, vertigo, anxiety, irritability, insomnia, loss of libido. B: Scanty or postponed menses; C: excessive menses; and D: irregular menses and intermenstrual bleeding.

Type of complaint	Country	Contraceptive method NORPLANT			IUD			Rate-ratio
		All women	With complaints	%	All women	With complaints	%	
A.	Chile	246	26	10.6	198	4	2.0	5.2
	China	227	15	6.6	194	4	2.1	3.2
	Sri Lanka	200	7	3.5	82	1	1.2	2.9
	Thailand	382	6	1.6	333	1	0.3	5.2
	All	1,055	54	5.1	807	7	0.9	5.7
	Adjusted rate-ratio							4.1
	confidence limits							2.1-8.0
	x^2							0.6
B.	Chile		21	8.5		8	4.0	2.1
	China		14	6.2		1	0.5	12.0
	Sri Lanka		38	19.0		2	2.4	7.8
	Thailand		33	8.6		3	0.9	9.6
	All		106	10.0		14	1.7	5.9
	Adjusted rate-ratio							4.4
	confidence limits							2.4-7.8
	x^2							6.3
C.	Chile		10	4.1		15	7.6	0.5
	China		6	2.6		23	11.9	0.2
	Sri Lanka		30	15.0		4	4.9	3.1
	Thailand		10	2.6		1	0.3	8.7
	All		55	5.2		19	2.4	2.1
	Adjusted rate-ratio							0.7
	confidence limits							0.4-1.2
	x^2							20.1[***]
D.	Chile		23	9.3		2	1.0	9.3
	China		89	39.2		20	10.3	3.8
	Sri Lanka		58	29.0		7	8.5	3.4
	Thailand		32	8.4		2	0.6	13.9
	All		202	19.1		31	3.8	5.0
	Adjusted rate-ratio							4.3
	confidence limits							3.0-6.3
	x^2							4.3

complaints utilized in this study is similar to that of some other studies (ICCR 1978, Du et al 1990) although we recorded complaints at unscheduled visits and not only at visits scheduled for the purpose of the study. This may explain the relatively high rates of complaints for both Norplant[R] and IUD acceptors. This client-oriented recording of complaints and the reasons for the health care system encounter, contrasts to the recording of complaints and health problems according to the opinion and attitudes of health care

Table 7A. Number of women with various complaints and total number of women by contraceptive method, country and body weight, and crude and weighted rate ratios (r-r). Complaints are: A: headache, weakness, nausea, vertigo, anxiety, irritability, insomnia and loss of libido; B: Scanty or postponed menses.

a Complaint A:

Country	Method	Weight			
		<50	50-49	60-69	>69
Chile	Norplant	11/ 39	25/128	11/ 58	4/ 21
	IUD	1/ 11	7/ 92	2/ 66	1/ 29
China	Norplant	8/ 45	11/114	12/ 62	0/ 6
	IUD	1/ 35	4/100	1/ 50	0/ 9
Sri Lanka	Norplant	15/176	1/ 20	1/ 4	0/ 0
	IUD	1/ 58	1/ 19	1/ 4	0/ 1
Thailand	Norplant	8/130	7/178	4/ 63	0/ 11
	IUD	3/114	1/148	1/ 52	1/ 19
Total	Norplant	42/390	44/440	28/187	4/ 38
	IUD	6/218	13/359	5/172	2/ 58
Crude	r-r	3.9(1.7-9.2)	2.8(1.5-5.1)	5.2(2.0-13.3)	
Weighted	r-r	3.4(1.0-8.2)	2.6(1.4-4.8)	4.8(1.8-13.0)	
x^2		0.77	1.09	1.92	

Complaint B:

Country	Method	Weight			
Chile	Norplant	9/ 39	39/128	15/ 58	4/ 21
	IUD	1/ 11	6/ 92	4/ 66	3/ 29
China	Norplant	9/ 45	7/114	4/ 62	0/ 6
	IUD	1/ 35	1/100	0/ 50	0/ 9
Sri Lanka	Norplant	111/176	11/ 20	2/ 4	0/ 0
	IUD	8/ 58	4/ 19	1/ 4	0/ 1
Thailand	Norplant	27/130	41/178	13/ 63	3/ 11
	IUD	13/114	19/148	4/ 52	1/ 19
Total	Norplant	156/390	98/440	34/187	7/ 38
	IUD	23/218	30/359	9/172	4/ 58
Crude	rr	3.8(2.5-5.9)	2.7(1.8-4.0)	3.5(1.7-7.2)	2.7(0.8-9.1)
Weighted	rr	2.9(1.8-4.6)	2.5(1.6-3.8)	3.2(1.5-6.8)	2.5(0.7-8.8)
x^2		4.16	4.17	0.50	0.56

providers which is typical for most studies on the side-effects of contraceptive drugs and devices. In addition to the differences among countries, it should be noted that although similar training was provided to thin vestigators in the different countries with regard to eliciting and probing complaints, there will obviously be in the interaction between health care provider and client, room for different and country-specific approaches to obtaining information on complaints and recording them. It is likely that clinic and country-specific factors have contributed to the relatively large differences in the rates of complaints among countries seen in this material. Therefore, unless adjustment is made for these factors, any population or ethnicity-specific pharmacological effect of a drug such as levonorgestrel may be difficult to detect and document since other factors mentioned above may confound possible affects attributable to drug metabolism or bioavailability. In this exploratory study complaints made by IUD users were regarded as a reference value of the level of complaints of the studied populations. In order to adjust for the effects of country and clinics on rates

Table 7B. Number of women with various complaints and total number of women by contraceptive method, country and body weight, and crude and weighted rate ratios (r-r). Complaints are: C: excessive menses; and D: irregular menses and intermenstrual bleeding.

b Complaint C:

Country	Method	<50	Weight 50-49	60-69	>69
Chile	Norplant	5/ 39	6/128	4/ 58	0/ 21
	IUD	1/ 11	10/ 92	4/ 66	2/ 29
China	Norplant	2/ 45	1/114	5/ 62	0/ 6
	IUD	7/ 35	12/100	5/ 50	2/ 9
Sri Lanka	Norplant	42/176	4/ 20	0/ 4	0/ 0
	IUD	7/ 58	6/ 19	1/ 4	0/ 1
Thailand	Norplant	3/130	7/178	1/ 63	1/ 11
	IUD	1/114	1/148	0/ 52	1/ 19
Total	Norplant	52/390	18/440	10/187	1/ 38
	IUD	16/218	29/359	10/172	5/ 58
Crude	r-r	1.8(1.0-3.2)	0.5(0.3-0.9)	0.9(0.4-2.2)	
Weighted	r-r	1.4(0.7-2.6)	0.5(0.3-1.1)	0.9(0.4-2.4)	
x^2_w		6.27	8.88*	0.13	

Complaint D:

Country	Method	<50	Weight 50-49	60-69	>69
Chile	Norplant	8/ 39	33/128	15/ 58	6/ 21
	IUD	2/ 11	7/ 92	5/ 66	1/ 29
China	Norplant	28/ 45	65/114	31/ 62	1/ 6
	IUD	6/ 35	20/100	8/ 50	2/ 9
Sri Lanka	Norplant	103/176	11/ 20	1/ 4	0/ 0
	IUD	12/ 58	3/ 19	0/ 4	0/ 1
Thailand	Norplant	18/130	30/178	8/ 63	1/ 11
	IUD	6/114	7/148	1/ 52	0/ 19
Total	Norplant	157/390	139/440	55/187	8/ 38
	IUD	26/218	37/359	14/172	3/ 58
Crude	r-r	3.4(2.2-5.1)	3.1(2.1-4.4)	3.6(2.0-6.5)	4.1(1.1-15.3)
Weighted	r-r	2.8(1.8-4.2)	3.1(2.2-4.5)	3.4(1.9-6.2)	2.9(0.6-14.2)
x^2		1.67	0.29	0.44	2.16

of complaints, the ratio of the rate of Norplant[R] and IUD users' complaints was used for among-country comparisons.

Using overall rates of complaints and overall method-related complaints as defined, there were no important differences in the rate-ratios, and tests of heterogeneity were statistically not significant. When method-related disturbances were further categorized into various types of bleeding disturbances, complaints of excessive menses were found to be statistically significant in the heterogeneity test (**Tables 5 and 6**). The country differing from the others in respect to this complaint was China, where substantially more IUD users than Norplant[R] users made this complaint. This pattern of complaints may find an explanation in the fact that the IUD used in Chinese centres is the "steel ring," which is reported to be associated with heavy bleeding during the first year of use.

Table 7C. Ratio of rate of Norplant^R and IUD users with various complaints by country and body weight, according to Tables 7A and B. Complaints are: A: headache, weakness, nausea, vertigo, anxiety, irritability, insomnia and loss of libido; B: Scanty or postponed menses; C: excessive menses; and D: irregular menses and intermenstrual bleeding.

c Type of complaint	Country	Weight			
		<50	50-59	60-69	>69
A.	Chile	3.1	2.6	6.3	5.5
	China	6.2	2.4	9.7	*
	Sri Lanka	4.9	1.0	1.0	*
	Thailand	2.3	5.8	3.3	*
	All	3.9	2.8	5.2	3.1
B.	Chile	2.5	4.7	4.3	1.8
	China	7.0	6.1	*	*
	Sri Lanka	4.6	2.6	2.0	*
	Thailand	1.8	1.8	2.7	5.2
	All	3.8	2.7	3.5	2.7
C.	Chile	1.4	0.4	1.1	*
	China	0.2	0.1	0.8	*
	Sri Lanka	2.0	0.6	*	*
	Thailand	2.6	5.8	*	1.7
	All	1.8	0.5	0.9	0.3
D.	Chile	1.1	3.4	3.4	8.4
	China	3.6	2.9	3.1	0.8
	Sri Lanka	2.8	3.5	*	*
	Thailand	2.6	3.6	6.6	*
	All	3.4	3.1	3.6	4.1

* Too few observations

The body weight-specific rate and rate-ratios were also analyzed, since body weight has been implicated in pregnancy rates for Norplant^R. Except for complaints of excessive menses as mentioned above, none of the tests of heterogeneity was statistically significant. Since an association between bodyweight and bleeding problems for users of Norplant^R has been observed, the rate-ratios were tabulated also across weight categories. We could not demonstrate any consistent trends of the rate-ratios.

We were unable in this material to find support for our working hypothesis that there are country-specific (or ethnic-specific) rates of complaints possibly attributable to differences of drug metabolism. However, this study should be seen as exploratory and has several limitations. The material available for analysis is small and the period of observation short. The restriction of the analysis to women who continued use of the method through the first year may have led to exclusion of women who discontinued the method because of more severe complaints. Also, data from only four of the participating countries were available for analyses. It is intended to further analyze the pattern of complaints in different countries in larger groups from the "Post-Marketing Surveillance of Norplant^R" and, of course, side effects with use of other endpoints such as the reason for discontinuation and diagnosis given by investigators.

Further to the use of complaints as an endpoint, it may be noted that complaints as perceived by the women and recorded as was done in this study may give important information about the content of counselling for the various methods. The consistently higher complaint rate of women recorded among Norplant^R acceptors compared to IUD users may be real, but may also be attributable to the fact that Norplant^R is a relatively recently introduced method and counsellors and other personnel may have been less experienced in counselling about Norplant^R, compared to the IUD.

References

1.   Cruikshank, J.K. & Beevers, P.G. *Ethnic factor in health and disease.* Wright. London 1989.
2.   Du M.K. et. al. Study of Norplant^R implants in Shanghai: three-year experience. *Int. J. Gynecol. Obstet.* 1990;33:345-357.
3.   Eisalo, A., Jarvinen, P.A., Luukainen, T. Liver-function tests during intake of contraceptive tablets in pre-menopausal women. *Brit Med J.* 1965; i:1416-7.
4.   International Committee for Contraceptive Research (ICCR). Contraception with long-acting subdermal implants: the measured and perceived effects international clinical trials. *Contraception* 1978;18:335-53.
5.   Lamberts, H., Wood, M. (Eds). *International Classification of Primary Care (ICPC).* Oxford University Press. 1988.
6.   Larsson-Cohn, U. Oral contraception and liver function tests. *Brit Med J.* 1965; i:1414-5.
7.   Sivin I. Clinical effects of Norplant^R subdermal implants for contraception. In (Mishell D.R. ed.) *Long-Acting Steroid Contraception* (pp. 89-116) Raven Press, New York 1983.
8.   Sivin, I. International experience with Norplant^R and Norplant^R-2 contraceptives. *Stud Fam Plan* 1988; 19; 81-94.
9.   Swyer, G.I.M., Little, V. Absence of hepatic impairment in long-term oral contraceptive users. *Brit Med J.* 1965; i:1412-4.
10.   WHO Special Programme of Research, Development and Research Training in Human Reproduction. Facts about an implantable contraceptive. *Bull WHO*, 1985; 63; 485-494.

REGIONAL AND INDIVIDUAL VARIATION IN BLEEDING PATTERNS ASSOCIATED WITH STEROID CONTRACEPTION

Elizabeth M. Belsey[1]

1. INTRODUCTION

The disturbance in menstrual bleeding induced by many methods of contraception is an important side-effect because of its potential impact on acceptability. Amenorrhea tends to be rejected because most women value menstruation as an assurance that they are not pregnant; it is also indicative of femininity, fertility and continuing youth, and widely viewed as important to general well-being[1]. Prolonged or unpredictable bleeding may, at best, be merely inconvenient or embarrassing, but in some cultures the taboos associated with menstruation can make any disruption to the normal pattern almost intolerable. As many as 25% of women recruited to trials of the long-acting injectable, depot medroxyprogesterone acetate (DMPA), discontinue method use within a year because of bleeding problems.[2,3] In recent studies of newer methods, the one-year cumulative discontinuation rates for bleeding irregularities have ranged from 9% among women using monthly injectables to 17% in a group using a vaginal ring.[4,5]

Although information on menstrual patterns is routinely obtained from women participating in contraceptive trials, attempts to describe and compare bleeding patterns have been hindered by inconsistencies in methods of data collection and difficulties in data analysis. Reports of clinical trials usually go no further than a comparison of the bleeding patterns experienced by women using different dosages or formulations of the same type of contraceptive. Information on the relationships between demographic variables,

E. BELSEY • World Health Organization, 1211 Geneva 27, Switzerland

Steroid Contraceptives and Women's Response,
Edited by R. Snow and P. Hall, Plenum Press, New York, 1994

individual characteristics and bleeding patterns among users of hormonal contraception could, however, be of immense value. The identification of sub-groups of women who would be likely to suffer more, or less, disruption to their bleeding patterns than the "norm" if they used a particular method of contraception would be helpful to clinicians in counselling the individual acceptor. On a different level, it would be useful to those responsible for the provision of family planning services at the national level. It might also provide some clues as to the basic mechanisms by which hormonal contraceptives affect menstrual bleeding, and thus point to the development of new formulations better suited to particular cultural or ethnic groups.

2. DEFINITIONS

The following definitions, based on terminology introduced by WHO in 1975[6] and clarified in 1986[7], have gained general acceptance and are used wherever possible:

Bleeding day: a day on which blood loss requiring the use of menstrual protection occurs.

Spotting day: a day on which blood loss insufficient to require protection occurs

Bleeding/spotting episode: any set of one or more consecutive bleeding or spotting days bounded at each end by bleeding-free days (A bleeding-free day is one on which neither bleeding nor spotting occurs.)

Bleeding-free interval: any set of one or more consecutive bleeding-free days bounded by bleeding or spotting days.

Bleeding/spotting segment: a bleeding/spotting episode and the immediately following bleeding-free interval.

Spotting episodes may also be defined and considered. Depending on the individual woman's definition of end-points for a "day," a one-day bleeding-free interval between two bleeding/spotting episodes may or may not be recorded. A bleeding-free interval is therefore sometimes redefined as any set of two or more consecutive bleeding-free days. A single bleeding-free day is then treated as part of the episode surrounding it.

The word "cycle" is deliberately avoided since it is "widely used and contains different and often undefined meanings."[6]

3. REGIONAL VARIATIONS IN BLEEDING PATTERNS

3.1 Review

Research on inter-regional differences in bleeding patterns has concentrated on non-contracepting women. Snowden and Christian[1] found that the number of days on which blood loss was recorded during a 90-day period varied from 12 in Mexico, 12 - 13 in India (depending on caste), 14 in Egypt, and 18 in England. Moreover, 36% of women in Egypt, 33% in India and 42% in Mexico estimated that their most recent bleeding/spotting episode had lasted less than 4 days. Only 18% of women in England, 16% in former Yugoslavia and 11% in Jamaica reported durations this short.

A comparison of bleeding patterns among women using a natural method of contraception in Europe, South East (SE) Asia, India/Pakistan and Latin America showed that European women had longer bleeding/spotting episodes (5.9 days, on average), shorter bleeding-free intervals (24.6 days) and thus more bleeding/spotting days (18 in every 90) than women in any other region.[8] SE Asian subjects had highly predictable, intermediate length episodes and intervals. Indian and Pakistani women reported more variable bleeding-free intervals than the other three groups, with a median range of 13 days over a one-year period. Latin American women had the shortest episodes and longest intervals (4.0 and 26.3 days, respectively). All these women were selected for their menstrual regularity, however, and the inter-regional differences observed among them may not be extrapolable to the wider population of non-contracepting women.

Variations in the preparations tested, and in methods of analyzing and reporting disturbances in bleeding patterns, are likely to impair the validity of any conclusions drawn by comparing the results of contraceptive trials conducted in different countries. The majority of oral contraceptive trials were conducted in the early 1970s, when it was still normal practice to use the cycle as the unit of analysis. Reports of such trials, however, may include the mean cycle length[9-12], the distribution of cycle lengths[9,11-15], or the change in cycle length since starting treatment[16]. A "normal cycle" has been variously defined as one lasting 25-32, 25-33, 25-35, or 26-30 days.[9,11,13-15] Amenorrhea has been described as no bleeding for one "cycle," for more than 45 days, or for more than 60 days.[9,10,13,14,16] The percentages of cycles with "intermenstrual" or "breakthrough" bleeding or spotting are usually given, but these terms are rarely defined. Moreover, summary statistics may be presented for each ordinal cycle, selected cycles, groups of cycles, or all cycles combined.

Similar problems occur in attempting to compare the menstrual experience of users of the subdermal implant, Norplant[R]. In publications summarizing the results of a multi-center trial conducted in Brazil, Chile, the Dominican Republic, Jamaica, Finland and Denmark, either the data from all centers are combined[17-21], or the discontinuation rate for menstrual problems is used as a surrogate measure of menstrual disturbance.[22] (This rate is subject to considerable bias: it is susceptible not only to cultural influence, but also to that of the woman's physician.) In the reports from individual countries, the statistics presented differ. Published results from the Dominican Republic include the percentages of women who recorded 5 or more episodes, 21 or more bleeding days, 5 or more spotting days, an episode exceeding 8 days or an interval exceeding 63 days per 90-day period[23]. The final report of the Chilean study gives only the mean numbers of bleeding, spotting and bleeding/spotting days in selected 90-day periods[24]. More recent single-country studies have been conducted in Egypt, Indonesia and Thailand. The Egyptian investigators presented the distributions of the number of bleeding episodes over the four 90-day periods in the first year of use, the proportions of women who reported changes in menstruation (categorized as frequent irregular, heavy or prolonged bleeding, spotting, amenorrhea or oligohypomenorrhea), and the distributions of the lengths of "intermenstrual bleeding" and spotting episodes[25]. The Thais also presented the proportions of women who reported menstrual changes, but used somewhat different categories (decreased or increased bleeding, more frequent bleeding, spotting, amenorrhea or irregular bleeding)[26]; while the Indonesians gave the percentages of women who complained of irregular bleeding and spotting or amenorrhea.[27]

The World Health Organization's Special Programme of Research, Development and Research Training in Human Reproduction is the only agency to have published direct comparisons between regions among women using each of a variety of contraceptives. An analysis of the menstrual diaries accumulated in a series of clinical trials conducted by the Programme was reported in 1988[8]. The diaries were completed by women using one of four types of hormonal contraceptive: combined oral pills containing either norethisterone acetate (NA) 1.0 mg + ethinyl estradiol (EE) 0.05 mg, levonorgestrel (LNG) 0.25 mg + EE 0.05 mg or LNG 0.15 mg + EE 0.03 mg; progestogen-only pills containing either NA 0.35 mg or LNG 0.03 mg; a vaginal ring releasing 20 ug of LNG daily; or the long-acting injectable, depot medroxyprogesterone acetate (DMPA), given every three months in dosages of 100 mg or 150 mg. The trials were conducted in a total of eight regions: Europe, SE Asia, China, India and Pakistan, Latin America, North Africa, Africa and the Caribbean.

Some of the differences between regions were consistent across contraceptive methods, while others were not. European women tended to record more bleeding/ spotting days than women in other regions, irrespective of the contraceptive method they used; Latin American women had relatively short episodes and long intervals, whether they were using combined pills or a vaginal ring. On the other hand, women using combined pills in India or Pakistan had fewer spotting episodes than women using the same method elsewhere; those using progestogen-only pills had more. SE Asian women given combined pills had short episodes and regular patterns; those using DMPA had long episondes and unpredictable patterns. Regional variations in bleeding patterns were particularly marked among women using DMPA, and increased over time: by their fourth injection interval, 25% of European women had amenorrhea, as compared with 72% of subjects in North Africa.

3.2. RECENT RESEARCH

For consistency with other papers published in the same series, the work described above was restricted to women who had used their contraceptive method for at least a year. This results in a considerable loss of information, since women who discontinue method use for menstrual reasons are precisely those with the worst bleeding disturbances[28]. The analyses have now been extended to women who continued method use (and completed a menstrual diary) for 90 days or more, and to a group using one of two monthly injectables, containing either depot medroxyprogesterone acetate 25.0 mg + estradiol cypionate 5.0 mg or norethisterone enanthate 50.0 mg + estradiol valerate 5.0 mg. However, since progestogen-only oral contraceptives were studied in only two regions, they have not been included. Details of the methods of analysis are given in the Appendix.

Table 1. Numbers of women in each region.

	Combined OC	Monthly injectables	Vaginal ring	DMPA
Europe	401	328	143	282
South East Asia	402	443	1	368
China	0	0	103	0
India/Pakistan	521	163	86	181
Latin America	379	724	151	0
North Africa	0	164	31	179
Africa	172	0	31	0
Caribbean	0	0	0	99
Total	1875	1822	546	1109

A total of 1875 combined oral contraceptive (OC) users, 1822 women given a monthly injectable 546 women using a vaginal ring and 1109 women given DMPA each completed a menstrual diary for at least 90 days. Their distribution by region is shown in Table 1. Figure 1 provides an overall comparison of the menstrual patterns recorded by women in the four groups. The four indices presented are those which most succinctly typify a pattern[29], since they describe the frequency, amount and predictability of bleeding. Women using a combined OC had the shortest bleeding/spotting episodes (4.1 days), and the most regular patterns. Users of the monthly injectables had fewer but longer episodes (5.1 days on average), and substantially reduced ability to predict the onset of their next episode: 50% of women experienced a difference of at least 24 days between their shortest and longest bleeding-free intervals. Vaginal ring users had the most frequent bleeding and shortest bleeding-free intervals (20.6 days) of the four groups, and comparable variability to that among women using the monthly injectables. Women given three-monthly injections of DMPA had chaotic bleeding patterns, with few, long episodes (1.5 every 90 days, with an average duration of 6 days), and long bleeding-free intervals (27.4 days). The range of interval lengths in this group had a median value of 51 days, and 25% of women had a range of 80 days or more.

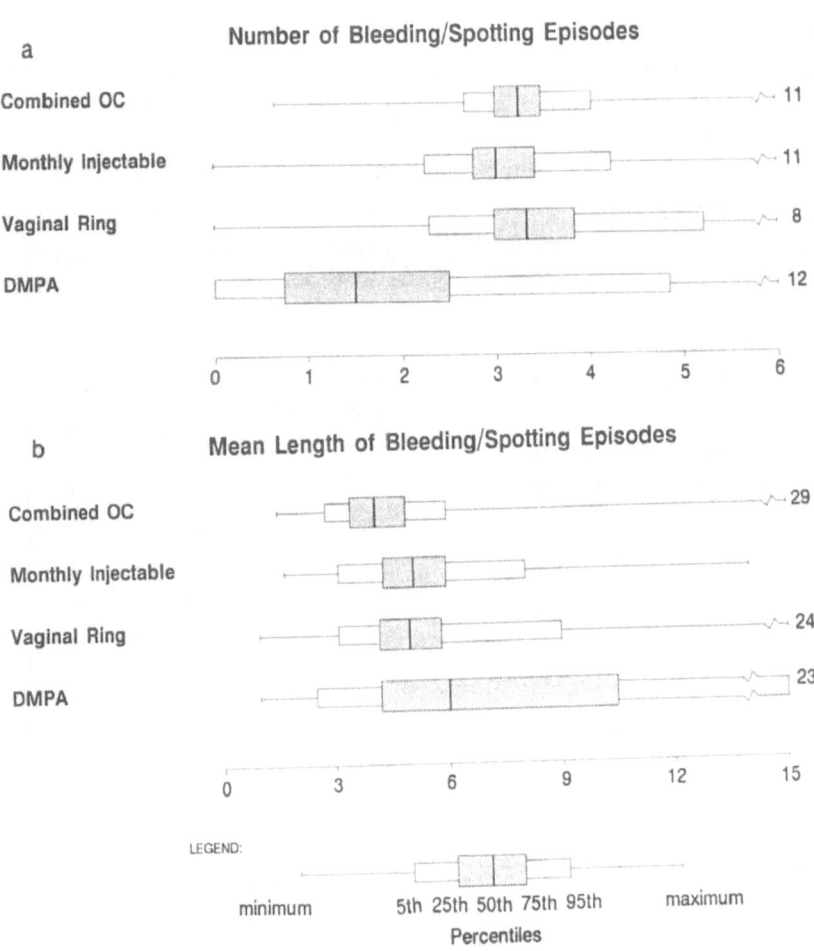

Figure 1A & B. Bleeding patterns by contraceptive method.

3.2.a Combined oral contraceptives

Trials of combined oral pills were conducted in five regions. In comparison with subjects elsewhere, European women had long bleeding/spotting episodes (4.6 days),

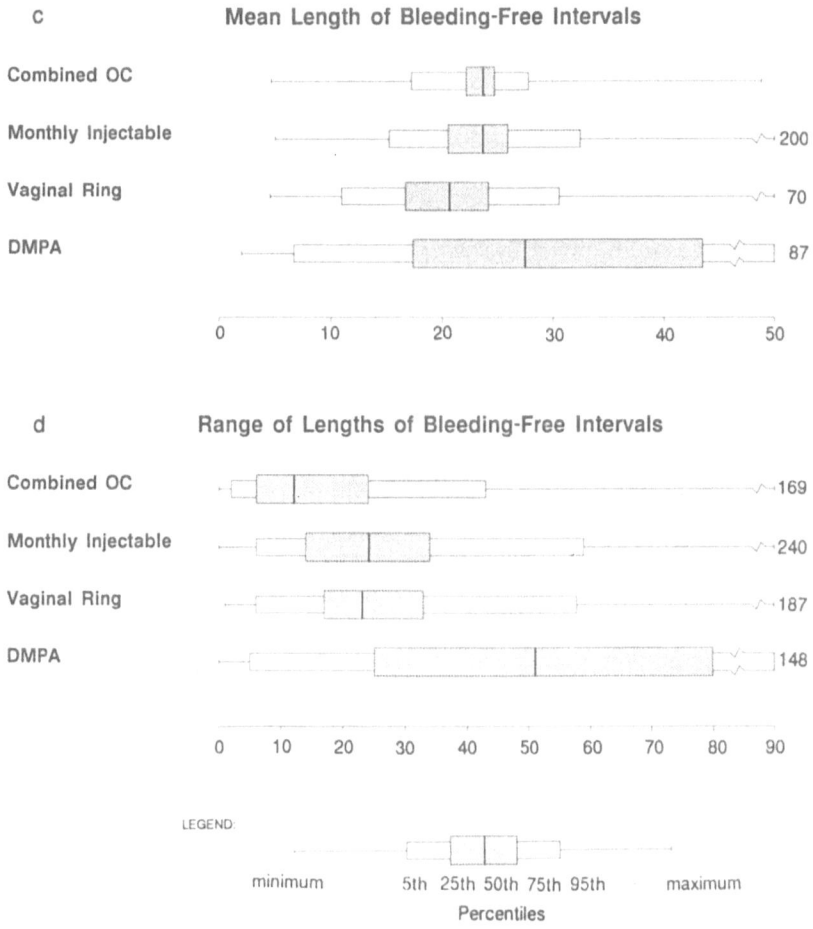

Figure 1C & D. Bleeding patterns by contraceptive method.

short bleeding-free intervals (22.8 days) and relatively unpredictable patterns (Figure 2). In contrast, SE Asian women recorded short episodes (3.7 days), long intervals (24.0 days) and more regular bleeding than the norm. Latin American women had similarly short episodes and long intervals; however, they also had the most variable patterns, with 50% of women recording a difference of at least 19 days between their shortest and longest bleeding-free intervals. African women had fewer but longer bleeding/spotting episodes than women in the other four regions.   Indian and Pakistani women reported intermediate bleeding patterns.

3.2.b Monthly injectables

Like the oral pills, the once-a-month injectables contain both estrogen and progestogen. Trials of these preparations were conducted in Europe, SE Asia, India and Pakistan, Latin America and North Africa. Figure 3 shows that European women had the most frequent bleeding (3.3 episodes every 90 days), longest bleeding/spotting episodes (averaging 5.3

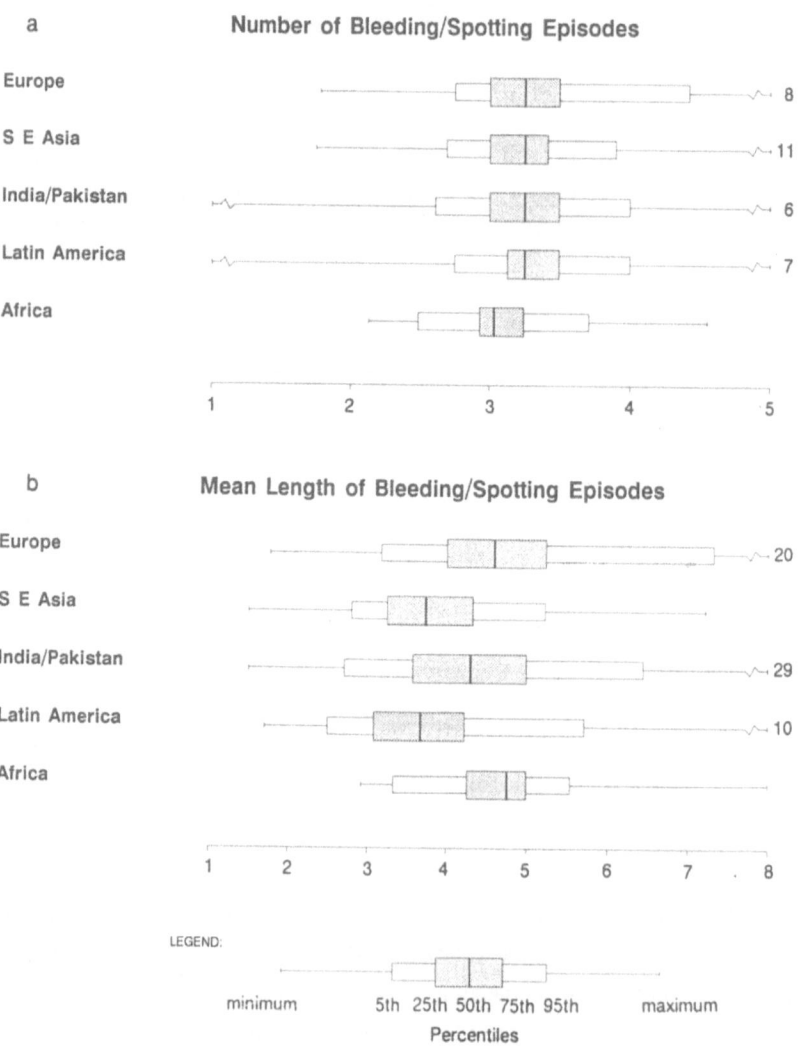

Figure 2A & B. Bleeding patterns by region-combined OC.

days), and shortest bleeding-free intervals (22.3 days). SE Asian women experienced some of the most variable bleeding patterns: the range of lengths of bleeding-free intervals had a median value of 26 days, and 25% of women had intervals whose lengths differed by 38 days or more. Indian and Pakistani women, like the Europeans, had long bleeding/spotting episodes, averaging 5.3 days. Women in Latin America recorded relatively frequent episodes of bleeding (3.1 every 90 days), interspersed by short (22.7 days), variable intervals. North African women had the fewest (2.8), shortest bleeding/spotting episodes (4.5 days), and the longest bleeding-free intervals (25.6 days). They also had the most predictable bleeding patterns: the median range of interval lengths was only 13 days, and less than 5% of women had intervals which differed by 35 days or more.

3.2.c Vaginal ring

There were six regions in which at least 31 and as many as 151 subjects used a vaginal ring for 90 days or more. European women had more frequent bleeding (3.7 episodes per 90-day period) and shorter bleeding-free intervals (18.6 days) than women in any other region (Figure 4). Chinese subjects recorded exceptionally long episodes (5.8 days) and short bleeding-free intervals (19.5 days). Women recruited in India/Pakistan had the shortest episodes (4.5 days) and relatively long, predictable bleeding-free intervals. Latin American and North African women had "average" patterns. Ring users in sub-Saharan Africa had the least frequent bleeding (3.0 episodes per 90-day period), and the longest (23.8 days), most variable bleeding-free intervals, with a range of lengths averaging 32 days.

3.2.d DMPA

DMPA was studied in five regions. Figure 5 shows that European women recorded the most frequent bleeding (2.9 episodes every 90 days), with relatively long episodes (7.0 days) and short bleeding-free intervals (26.3 days). SE Asian women had fewer episodes (2.3) and more variable patterns, with 25% of women experiencing an 80-day difference between their shortest and longest bleeding-free intervals. Indian and Pakistani women had longer bleeding/spotting episodes (7.3 days), and shorter bleeding-free intervals (25.8 days) than women in any other region. North African women were most subject to amenorrhea: 25% of women did not bleed at all during DMPA use, and the median number of bleeding/spotting episodes over each 90-day period was only 0.8. Women who did record any blood loss had short episodes (4.5 days). They had the longest but least variable bleeding-free intervals, with a median range of bleeding-free interval lengths of only 39 days, in comparison with 51 days overall. In contrast, the median range for Caribbean women was 74 days, and 25% of women had intervals which varied in length by as much as 122 days.

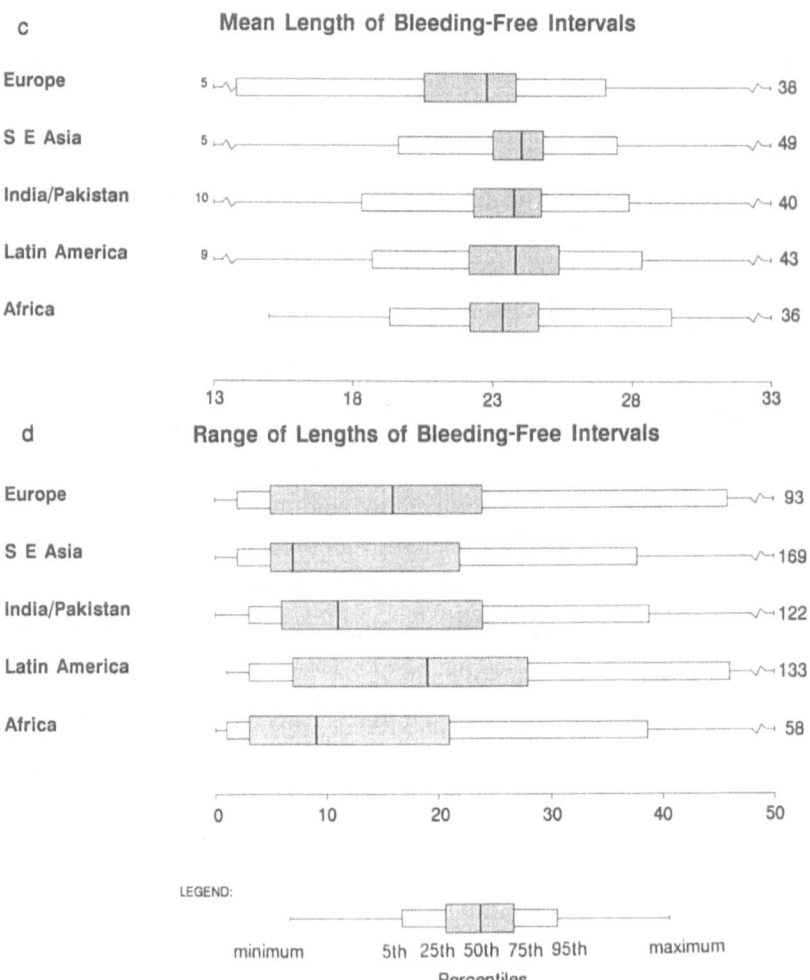

Figure 2C & D. Bleeding patterns by region-combined OC.

3.2.e Conclusions

Regardless of the contraceptive they used, European women had more frequent bleeding than women in most other regions; in consequence, they also tended to have the shortest bleeding-free intervals. SE Asian women using a combined pill had short episodes, and long, predictable intervals; when they were given a monthly or a three-monthly injectable, however, their bleeding became very variable. In the one study in which they were included, Chinese women had substantially longer bleeding/spotting episodes than any other group. Indian and Pakistani women given either the monthly or the three-monthly injectable had longer episodes than women studied elsewhere, while those given a vaginal ring had the shortest. Latin American women using a combined pill had short episodes and long, variable intervals; those given a monthly injectable had normal length episodes but short, again variable intervals; while those given a vaginal ring had "average" bleeding patterns.

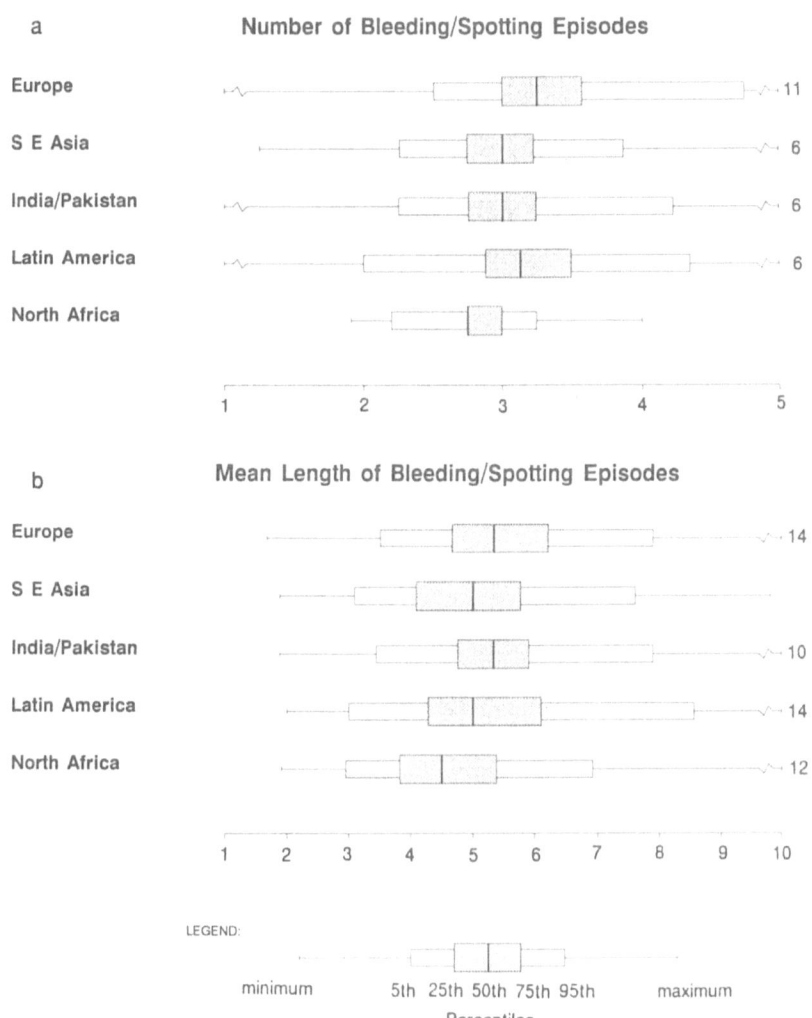

Figure 3A & B. Bleeding patterns by region - monthly injectable.

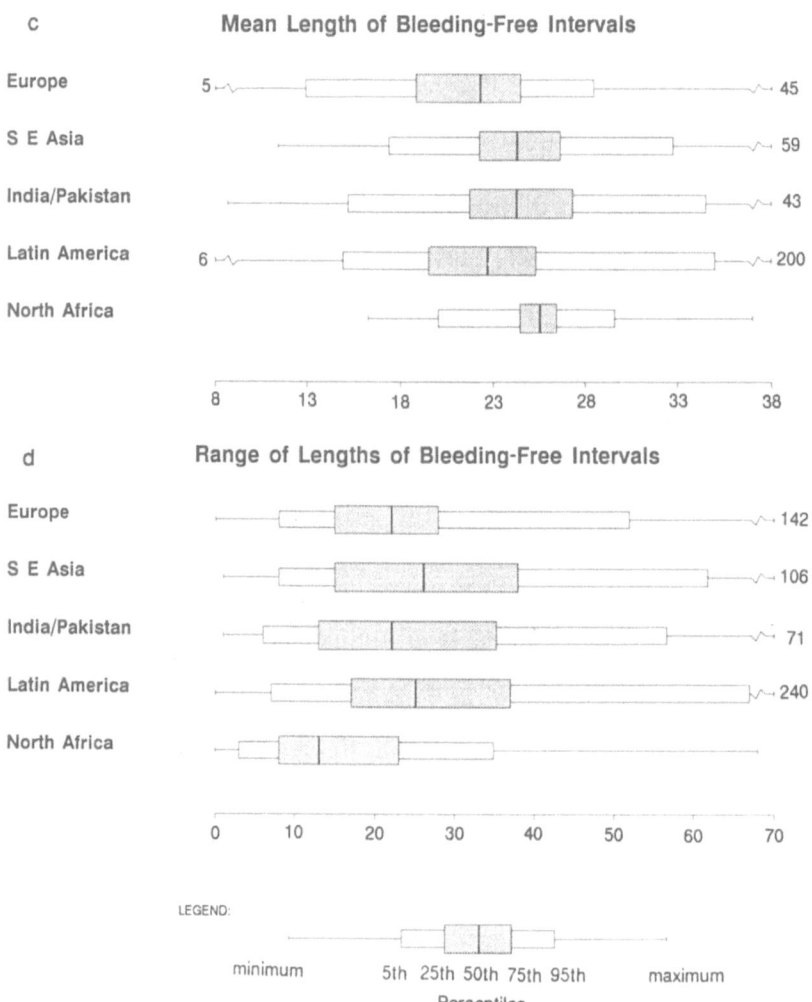

Figure 3C & D. Bleeding patterns by region - monthly injectable.

North African women given either of the two types of injectable had the shortest bleeding/spotting episodes and longest bleeding-free intervals; irrespective of method, women from this region also had the least variable bleeding patterns. African women were included only in the trials of combined OCs and the vaginal ring; in both cases they had infrequent bleeding and a tendency to long episodes. However, African women using the combined pill had predictable patterns, relative to women in other regions; those using the vaginal ring had the most variable patterns.

These findings raise at least as many questions as they answer. Perhaps the most

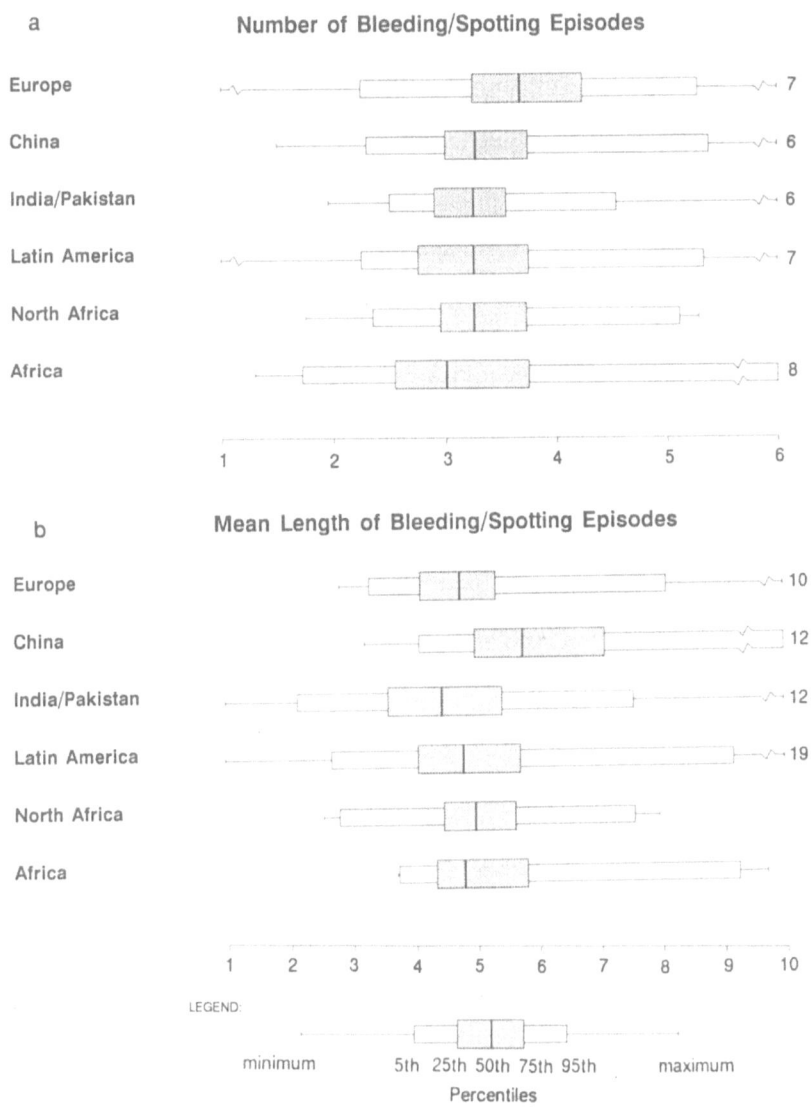

Figure 4A & B. Bleeding patterns by region - vaginal ring.

important of these concerns their validity. There are several factors other than genuine differences in bleeding patterns that could have influenced the results. It is unlikely that the quality of diary completion was the same in all regions. Snowden and Christian[1] had varying degrees of success in their attempts to obtain daily records of bleeding from samples of women in 11 cultural groups. They concluded that a woman's ability to complete a menstrual diary is largely determined by her literacy level, and that some women are simply incapable of keeping a daily record for themselves. On the other hand, all women included in these analyses were capable of completing a diary card for

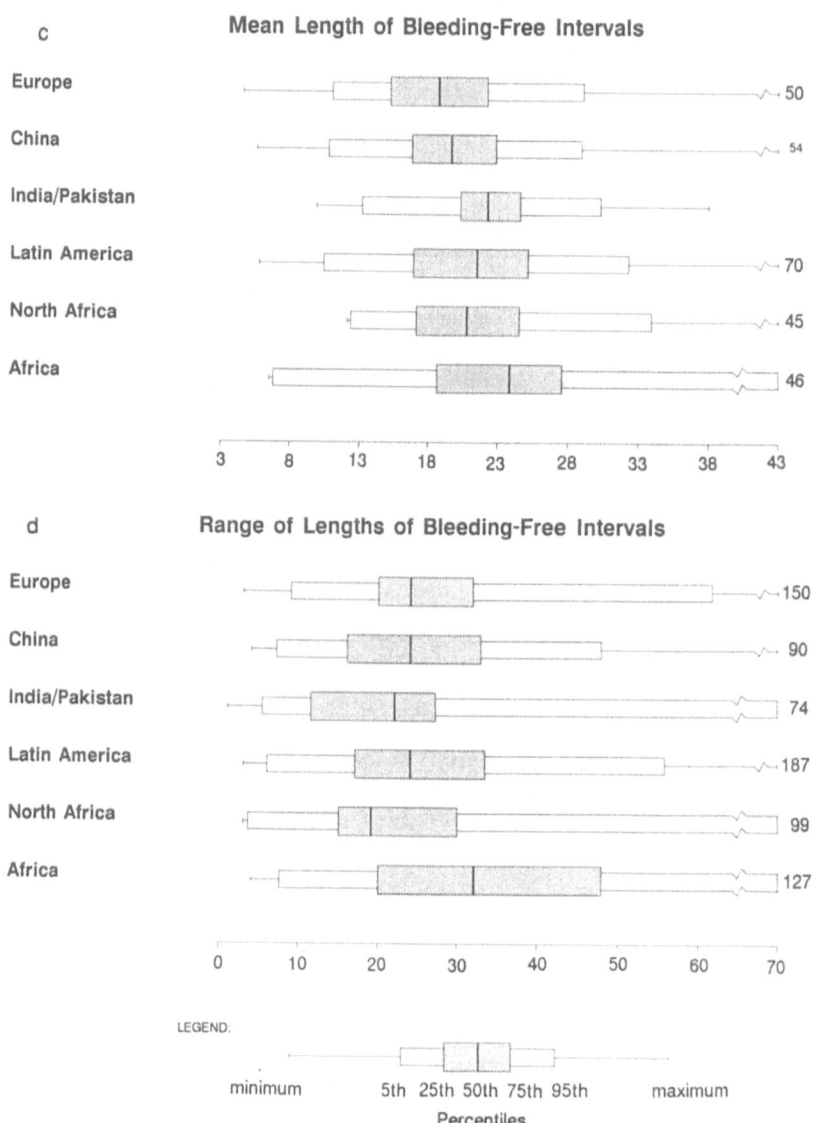

Figure 4C & D. Bleeding patterns by region - vaginal ring.

at least three months and, in many cases, up to a year, and sufficiently motivated to do so.

The prevalence of use of menstrual protection may also have affected the recording of blood loss. Snowden and Christian[1] found that protection was used almost universally by women in England, former Yugoslavia, Jamaica and Mexico, but 15% of women in the Philippines and 27% of Egyptian women reported using no protection at all. This may not have a dramatic effect on the recording of bleeding, but it reduces the usefulness of the concept of spotting.

Regional variations in the discontinuation rate, and in the relative importance of the menstrual and non-menstrual reasons for discontinuing method use, could also have influenced the results. Finally, a number of variables on which bleeding patterns may well depend, such as diet and physical activity, were not measured in any of the trials; nor, therefore, could any adjustment for their effects be made in the analysis.

The relative impact of contraceptive method on the inter-regional differences in bleeding patterns is difficult to assess. Some of the variation was clearly pre-existing. For example, European women using a natural method of contraception recorded shorter bleeding-free intervals than SE Asian, Indian/Pakistani or Latin American subjects[8]. Whatever hormonal method they used, they still tended to have shorter intervals than women in most other regions.

Nevertheless, the findings do seem to indicate that the route of administration of a contraceptive may have a greater impact than its content. There were more similarities between the effects of the two injectables, only one of which contains estrogen, than between those of the two combined contraceptives or the two progestogen-only preparations.

4. INFLUENCE OF INDIVIDUAL CHARACTERISTICS

4.1 Review

In healthy, non-contracepting women, segments grow shorter with increasing age to reach their minimum length in the mid to late thirties, and then lengthen again towards menopause[30,31]. The greatest variability in segment length is observed before the age of 20 and after the age of 40. In a sample of untreated women drawn from eight countries, higher parity was found to be associated with an increase in the total number of bleeding/spotting days per reference period, and in the variability of bleeding-free intervals[32]. Most studies which have investigated the effects of body mass index on menstruation outside the extreme of anorexia nervosa have found no relationship between body fat indices and segment length[33,34]. In a group of women using a natural method of contraception, however, a higher ponderal index was found to be associated with more frequent, shorter episodes of bleeding, and less variable bleeding-free intervals[35].

The literature contains few, mainly passing, references to the effects of individual characteristics on bleeding patterns in women using hormonal contraception. Tejuja *et al.* could find no relationship between weight and menstrual patterns among women using progestogen-only oral contraceptives[15]. Among users of norethisterone enanthate (NET-EN) injected every 12 weeks, women heavier than 62 kg were found to experience more bleeding and spotting days than women weighing less than 47 kg, but the average difference between the two groups was only 1.6 days[36]. No such difference was observed among DMPA users. Mayes and Harding failed to find any associations between age, age at menarche, weight, height, parity or number of abortions and a number of bleeding pattern indices among women using a combined OC, NET-EN or DMPA[32].

4.2 Recent research

The series of papers published in 1988 by the World Health Organization's Special Programme of Research, Development and Research Training in Human Reproduction included a description of the relationships between bleeding patterns and age, age at menarche, ponderal index, and obstetric and contraceptive history among women using combined or progestogen-only oral contraceptives, a vaginal ring or DMPA[35]. These analyses have now been extended to women who used their contraceptive method for at least 90 days, and to a group using monthly injectables. Table 2 shows selected characteristics of the subjects in each group. There was less than a two-year difference in mean age between users of the four types of contraceptive. Only 81 (4%) OC users were nulligravida, and all women recruited to trials of the vaginal ring or DMPA had been pregnant previously. The proportion of women whose last pregnancy had been terminated varied from 22% in the DMPA group to 35% of vaginal ring users. Between 60% and 85% of women in each group had used some method of contraception in the previous two years.

4.2.a Combined oral contraceptives

In the combined pill group, increasing age was associated with an increase in the frequency of bleeding and a decrease in the variability of bleeding-free intervals (Table 3). Subjects with a higher Quetelet Index had longer but less variable bleeding-free intervals. Mean episode length increased, and bleeding patterns became less predictable, with higher parity. Women whose last pregnancy had ended in abortion had shorter bleeding/spotting episodes than those who had had a live birth. The most influential variable in this group, however, was the time since the end of the woman's last pregnancy. Women who had been more recently pregnant had more frequent, longer and more variable episodes of bleeding, and shorter bleeding-free intervals. There were only scattered relationships with the most recent contraceptive method used: women who had used a method other than an oral pill or IUD had fewer episodes and thus longer bleeding-free intervals than women who had used a non-interfering method or none at all.

4.2.b Monthly injectables

Among users of the two monthly injectables, older women had less frequent bleeding, and thus longer bleeding-free intervals (Table 4). Heavier women also had fewer, shorter bleeding/spotting episodes, and longer, more variable bleeding-free intervals.

Table 2. Subject characteristics by method of contraception.

	Age (years)		Age at menarche (years)		Quetelet Index		Number of pregnancies		Months since end of last pregnancy	
	Mean	S.D.	Mean	S.D.	Mean	S.D.	Mean	S.D.	Mean	S.D.
Combined OC										
Europe	26.1	5.5	13.2	1.5	21.9	3.2	1.9	1.6	30.5	39.8
South East Asia	25.6	4.8	13.9	1.7	20.4	2.6	2.5	1.5	11.2	15.4
India/Pakistan	26.5	4.2	14.0	1.2	19.9	3.4	2.9	1.5	14.8	17.3
Latin America	27.2	5.1	13.4	1.6	22.5	3.3	3.1	1.9	33.4	35.9
Africa	26.3	5.0	14.7	1.5	22.6	3.8	3.9	2.2	14.8	10.7
Monthly injectables										
Europe	28.7	3.9			22.8	2.9				
South East Asia	26.6	3.9			20.3	3.2				
India/Pakistan	27.1	4.0			21.6	4.1				
Latin America	25.2	4.2			23.7	3.5				
North Africa	30.0	3.7			29.1	5.9				
Vaginal ring										
Europe	28.3	4.3			23.0	3.5	2.0	1.3	41.6	43.4
China	30.4	2.2			20.6	1.9	2.0	0.9	18.5	21.1
India/Pakistan	28.7	3.3			21.8	3.0	2.7	1.3	32.2	24.7
Latin America	26.4	4.0			23.4	3.5	2.4	1.4	29.3	30.8
North Africa	26.4	3.9			24.1	4.1	2.5	2.3	13.0	12.3
Africa	27.9	4.1			25.1	2.8	3.8	1.9	28.4	26.3
DMPA										
Europe	30.0	4.8	13.5	1.4	23.5	3.4	3.6	2.0	33.3	42.0
South East Asia	25.3	4.5	14.1	1.7	21.3	3.2	2.2	1.3	12.4	18.1
India/Pakistan	28.3	4.3	13.0	0.9	21.5	3.9	5.0	2.3	14.3	13.1
North Africa	30.7	4.0	11.7	0.8	28.3	5.6	4.5	1.7	29.4	29.5
Caribbean	24.2	4.3	13.3	1.4	22.2	3.6	2.2	1.2	25.2	18.6

Table 3. Associations between bleeding pattern indicies and individual characteristics (regression coefficient ± S.E.)in combined oral contraceptive users.

	+Number of B/S episodes	Mean episode length	Mean interval length	Range of interval lengths
Age (years)	+0.07 ± 0.03 *			-0.20 ± 0.10 *
+Quetelet Index			+0.59 ± 0.25 *	-2.27 ± 1.11 *
Number of pregnancies		+0.04 ± 0.02 *		+0.69 ± 0.26 **
Last pregnancy ended in abortion		-0.14 ± 0.07 *		
++Months since end last pregnancy	-1.33 ± 0.55 *	-0.44 ± 0.12 ***	+1.11 ± 0.35 **	
Used 'other' contraceptive method	-1.64 ± 0.66 *		+1.11 ± 0.42 **	

Coefficients multiplied by: +10 ++100
Significance of difference from 0: * $p < 0.05$ ** $p < 0.01$ *** $p < 0.001$

Use of an oral contraceptive within the previous two years had a similar effect: less frequent bleeding, shorter, less variable episodes and longer intervals. Women who had used a method other than an oral pill or IUD had fewer episodes than those who had used a barrier or natural method or none.

4.2.c Vaginal ring

In the vaginal ring group, a higher Quetelet Index was associated with longer bleeding-free intervals, as a result of a (non-significant) decrease in the length of each episode, rather than their frequency (Table 5). Women of higher parity were less able to predict the onset of their next bleeding/spotting episode, while those whose last pregnancy had ended in abortion recorded shorter bleeding-free intervals than women who had had a live birth. Previous oral contraceptive users had shorter episodes; women who had been using an IUD had less variable bleeding-free intervals.

4.2.d DMPA

Among DMPA users, bleeding patterns were unaffected by age or obstetric history. Quetelet Index, however, had a substantial impact: heavier women noted less frequent, shorter episodes, and longer bleeding-free intervals. [Coefficient $\pm$ S.E. (each multiplied by 10), significance $= -0.65 \pm 0.12$, $p<0.001$; -1.52 ± 0.77, $p<0.05$; $+6.84 \pm 2.67$, $p=0.01$, respectively]. Spotting-only episodes are an integral part of the bleeding pattern in this group. Women in whom menarche had occurred later had fewer spotting episodes and thus longer segments. Those who had taken oral contraceptives in the two years prior to DMPA use also had fewer spotting episodes and, in consequence, a lower range of lengths of all bleeding/spotting episodes.

4.2.e Conclusions

The associations found were often inconsistent or difficult to interpret. Age was related to bleeding patterns among women using the two "combined" hormonal contraceptives, the OCs and the monthly injectables. However, increasing age was associated with an increased frequency of bleeding among combined OC users, but a decreased frequency among women given one of the monthly injectables.

Quetelet Index was the characteristic with the greatest impact on bleeding. In all four groups, the more obese women had longer bleeding-free intervals. Heavier combined OC users also had more predictable intervals; in the monthly injectable group, greater obesity was associated with less frequent, shorter episodes of bleeding but less predictable intervals; in the DMPA group, it was again associated with fewer, shorter episodes, but had no effect on the variability of the pattern. Slimmer DMPA users thus had patterns closer to untreated

Table 4. Associations between bleeding pattern indices and individual characteristics (regression coefficient ± S.E.) in monthly injectable users.

	[+]Number of B/S episodes	Mean episode length	Mean interval length	Range of interval lengths
Age (years)	-0.14 ± 0.04 ***		+0.14 ± 0.05 **	
[+]Quetelet Index	-1.98 ± 0.42 ***	-0.19 ± 0.10 *	+2.01 ± 0.51 ***	+5.05 ± 1.25 ***
Used oral contraceptives	-0.87 ± 0.44 *	-0.35 ± 0.10 ***	+1.45 ± 0.53 **	
Used 'other' contraceptive method	-1.00 ± 0.51 *			

[+]Coefficients multiplied by 10
Significance of difference from 0: * $p < 0.05$ ** $p < 0.01$ *** $p < 0.001$

Table 5. Associations between bleeding pattern indices and individual characteristics (regression coefficient ± S.E.) in vaginal ring users.

	Mean episode length	Mean interval length	Range of interval lengths
+Quetelet Index		+2.28 ± 0.93 *	
Number of pregnancies			+1.89 ± 0.75 *
Last pregnancy ended in abortion		-1.39 ± 0.69 *	
Used oral contraceptives	-0.45 ± 0.23 *		
Used IUD			-6.15 ± 2.74 *

+Coefficients multiplied by 10
Significance of difference from 0: * p < 0.05   ** p < 0.01   *** p < 0.001

patterns than their heavier counterparts. This is not consistent with the observations of lower blood levels and reduced efficacy of other progestogen-only methods in heavier women[37].

Higher parity was associated with less predictable bleeding-free intervals among both combined OC and vaginal ring users. This effect was independent of age. Women using a combined OC whose last pregnancy had ended in abortion had shorter bleeding/spotting episodes than those who had had a live birth; women given a vaginal ring had shorter bleeding-free intervals if their last pregnancy had been terminated. Previous use of an oral contraceptive resulted in less frequent bleeding among monthly injectable, vaginal ring and DMPA users; previous use of a contraceptive method other than oral pills or an IUD had the same effect in the combined OC and monthly injectable groups. These results are particularly difficult to assess since the patterns observed reflect the joint effects of a previous and a current contraceptive method.

In the DMPA group, women who had a later menarche had fewer spotting-only episodes. This suggests that DMPA users who have a shorter menstrual experience may possibly be more prone to DMPA-induced amenorrhea. Previous studies on young non-contracepting women have shown a similar correlation between age at menarche and mean segment length[30,33].

Findings such as these can be statistically significant without being clinically meaningful. This is especially true in the combined OC group, where the predicted increase in the number of bleeding/spotting episodes per 90-day period is 0.007 for every year of age. Thus the predicted difference in the number of episodes between a 20 year-old and a 40 year-old woman is only 0.140. Similarly, the change in weight required to cause a one-day increase in the average length of bleeding-free intervals in a woman 1.65 m in height is almost 5 kg. In the other three groups, the regression coefficients tend to be numerically larger. Some effects on the length or variability of bleeding-free intervals, particularly of Quetelet Index among monthly injectable and DMPA users and previous use of an IUD among women given a vaginal ring, would be noticeable to the individual user.

The association between Quetelet Index and bleeding patterns merits further investigation. It appears, however, that the combined effects of contraceptive method and some undetermined aspect of region, as yet, are strong enough to overshadow all other factors. The question "What is region" remains to be explored in future studies.

5. APPENDIX

5.1 Clinical trials

The menstrual diary records were completed by 6522 women recruited to six clinical trials, which have been described in detail elsewhere[2,4,5,38,39]. The trials were conducted in a total of 39 centers. The number of centers in which each type of contraceptive was studied varied from seven to 17. The centers were divided into eight regions as follows: Europe, including the USSR, 9 centers; SE Asia, 5; China, 3; India and Pakistan, 4; Latin America, 13; North Africa, 2; Africa, 2; and the Caribbean, 1.

Age, weight, height and last contraceptive method used were recorded routinely in all six trials. In addition, number of previous pregnancies, outcome of the woman's last pregnancy and time since the end of this last pregnancy were noted for all women except those using a monthly injectable, and age at menarche was obtained from women using an oral contraceptive or DMPA.

5.2 Recording and summarization of bleeding patterns

All women were asked to complete a menstrual diary card for the duration of their trial, and to differentiate between bleeding days and spotting days. The reference period method[40] was used to analyze the records. Diaries which had been completed for less than 90 days were excluded. Analysis commenced on the first day of treatment and was limited to events occurring in the first 360 days of method use. The alternative definition of a bleeding-free interval, as any set of two or more consecutive bleeding-free days bounded by bleeding or spotting days, was used in these analyses. The following bleeding pattern indices were calculated for each woman for the single 360-day reference period: the numbers of bleeding/spotting days, spotting days, bleeding/spotting episodes begun and spotting episodes begun; and the mean and range of bleeding/spotting episode, bleeding-free

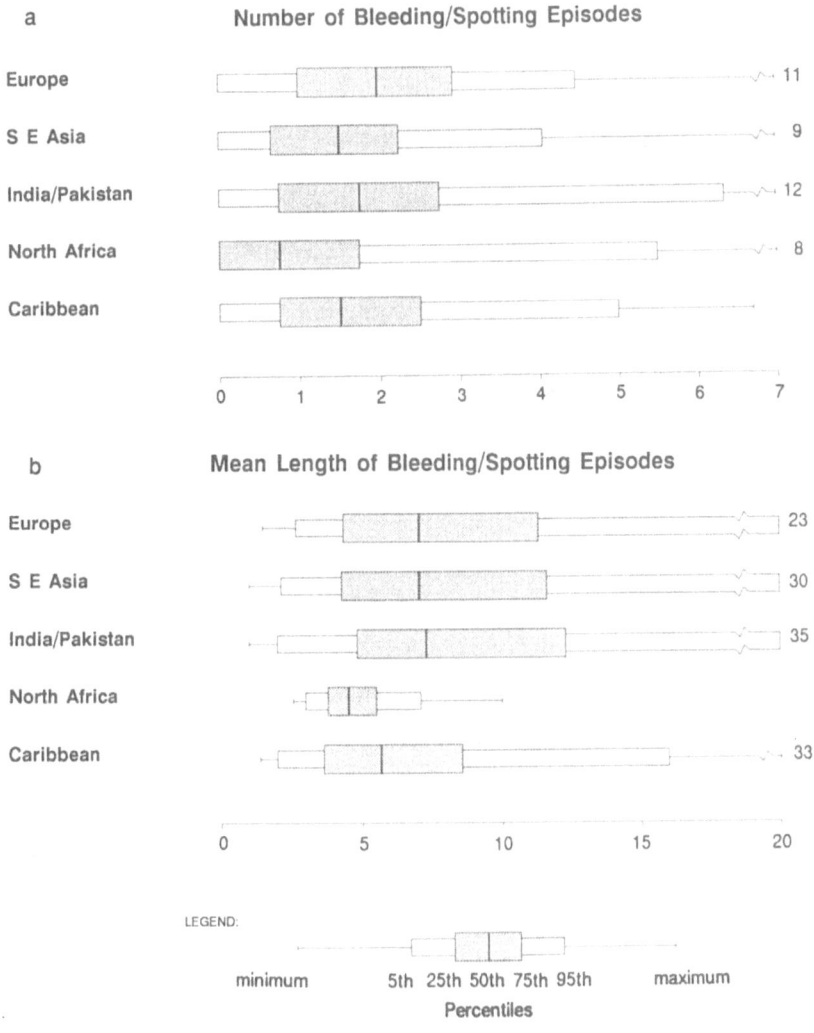

Figure 5A & B. Bleeding patterns by region - DMPA.

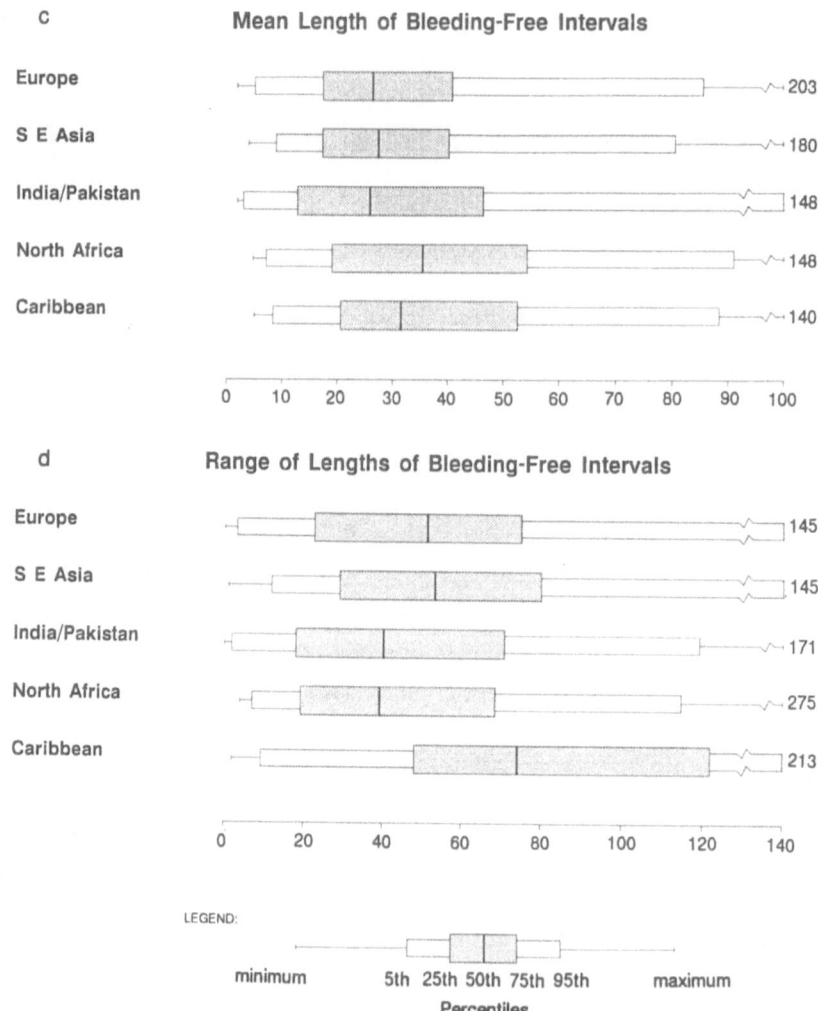

Figure 5C & D. Bleeding patterns by region - DMPA.

interval and bleeding/ spotting segment lengths. The counts of bleeding/spotting and spotting days and episodes were then adjusted to the standard 90-day reference period recommended by WHO[7] by dividing by the length of the diary and multiplying by 90.

5.3 Statistical methods

Preliminary results showed that bleeding patterns were more closely related to the woman's region of residence than to any other factor. The analysis was therefore conducted in two parts. In the first, bleeding patterns were compared across regions within each method group. For descriptive purposes, the indices were summarized over the subjects in each group/region by calculating the minimum value, the 5th, 25th, 50th, 75th and 95th percentiles and the maximum value. The statistical significance of differences in the indices between regions was determined by means of non-parametric analysis of variance[41]. In the second part, multiple regression analysis was used to examine the relationships between the bleeding pattern indices and age, age at menarche, Quetelet Index [weight (kg)/height^2 (m)], number of pregnancies, outcome of the woman's last pregnancy, time since the end of her last pregnancy, and last contraceptive method used. Pregnancy outcome and last contraceptive method used are categorical variables and were therefore replaced by dummy variables (coded 0 or 1) with live birth as the reference outcome and the combination of none/barrier/natural as the reference method of contraception. Since apparent "ethnic" differences between regions could have been caused by differences in other characteristics of the subjects studied, these regression analyses were also used to confirm the existence of independent regional variations in the bleeding pattern indices: region was included in the regression equations as a series of dummy variables, with Europe as the reference region.

References

1. Snowden R, Christian B, (eds). *Patterns and Perceptions of Menstruation.* Croom Helm, London. 1983.
2. World Health Organization Task Force on Long-Acting Systemic Agents for Fertility Regulation. A multicenterd Phase III comparative clinical trial of depot medroxyprogesterone acetate given three-monthly at doses of 100mg or 150mg: 1. Contraceptive efficacy and side effects. *Contraception* 34:223-235. 1986.
3. WHO Special Programme of Research, Development and Research Training in Human Reproduction: Task Force on Long-Acting Agents for the Regulation of Fertility. Multi-national comparative clinical trial of long-acting injectable contraceptives: norethisterone enanthate given in two dosage regimens and depot-medroxyprogesterone acetate. Final report. *Contraception* 28:1-20. 1983.
4. World Health Organization Task Force on Long-Acting Systemic Agents for Fertility Regulation. A multicenterd Phase III comparative study of two hormonal contraceptive preparations given once-a-month by intramuscular injection: I. Contraceptive efficacy and side effects. *Contraception* 37:1-20. 1988.
5. WHO Task Force on Long-Acting Systemic Agents for Fertility Regulation. Microdose intravaginal levonorgestrel contraception: a multicenter clinical trial. I. Contraceptive efficacy and side-effects. *Contraception* 41:105-124. 1990.

6. WHO Task Force on Acceptability of Fertility Regulating Methods (Section on Patterns and Perceptions of Menstrual Bleeding). Report on a meeting on the statistical analysis of menstrual bleeding data. Geneva, 27-29 October 1975, (unpublished).

7. Belsey EM, Machin D, d'Arcangues C. The analysis of vaginal bleeding patterns induced by fertility regulating methods. *Contraception* 34:253-260. 1986.

8. Belsey EM, Peregoudov S, Task Force on Long-Acting Systemic Agents for Fertility Regulation. Determinants of menstrual bleeding patterns among women using natural and hormonal methods of contraception. I. Regional variations. *Contraception* 38:227-242. 1988.

9. Arnt IC, *et. al.* Low-dose combination oral contraceptives: a controlled clinical study of three different norgestrel-ethinyl estradiol ratios. *Fertil Steril* 28:549-553. 1977.

10. Larranaga A, *et. al.* Clinical evaluation of two biphasic and one triphasic norgestral/ethinyl estradiol regimens. *Int J Fertil* 23:193-199. 1978.

11. Rubio B, Mischler TW, Berman E. Further experience with a once-a-month oral contraceptive: quinestrol-quingestanol. *Fertil Steril* 23:734-738. 1972.

12. Vessey MP, *et. al.* Randomised double-blind trial of four oral progestagen-only contraceptives. *Lancet* 1:915-922. 1972.

13. Apelo R, Veloso I. Clinical experience with ethinyl estradiol and d-norgestrel as an oral contraceptive. *Fertil Steril* 26:283-288. 1975.

14. Ferrari A, *et. al.* The menstrual cycle in women treated with d-norgestrel 37.5 micrograms in continuous administration. *Int J Fertil* 18:133-140. 1973.

15. Tejuja S, *et. al.* Experience with 50 mcg and 75 mcg dl-norgestrel as a mini-pill in India. *Contraception* 10:385-394. 1974.

16. Prsic J, Kicovic PM. Low-dose lynestrenol as a contraceptive method. *Contraception* 8:315-326. 1973.

17. The International Committee for Contraception Research of the Population Council. Contraception with long acting subdermal implants: I. An effective and acceptable modality in international clinical trials. *Contraception* 18:315-333. 1978.

18. The International Committee for Contraception Research of the Population Council. Contraception with long acting subdermal implants: II. Measured and perceived effects in international clinical trials. *Contraception* 18:335-353. 1978.

19. Faundes A, Sivin I, Stern J. Long acting contraceptive implants. An analysis of menstrual bleeding patterns. *Contraception* 18:355-365. 1978.

20. Sivin I, *et. al.* Three-year experience with NORPLANT™ subdermal contraception. *Fertil Steril* 39:799-808. 1983.

21. Sivin I, *et. al.* A four-year clinical study of NORPLANT[R] implants. *Stud Fam Plann* 14:184-191. 1983.

22. Sivin I, *et. al.* Norplant: reversible implant contraception. *Stud Fam Plann* 11:227-235. 1980.

23. Alvarez F, *et. al.* Comparative clinical trial of the progestins R-2323 and levonorgestrel administered by subdermal implants. *Contraception* 18:151-162. 1978.

24. Diaz S, *et. al.* A five-year clinical trial of levonorgestrel silastic implants (Norplant™). *Contraception* 25:447-456. 1982.

25. Shaaban MM, *et. al.* A prospective study of NORPLANT[R] implants and the TCu 380Ag IUD in Assiut, Egypt. *Stud Fam Plann* 14:163-169. 1983.

26. Satayapan S, Kanchanasinith K, Varakamin S. Perceptions and acceptability of NORPLANT[R] implants in Thailand. *Stud Fam Plann* 14:170-176. 1983.

27. Lubis F, *et. al.* One-year experience with NORPLANT[R] implants in Indonesia. *Stud Fam Plann* 14:181-184. 1983.

28. Belsey EM, Task Force on Long-Acting Systemic Agents for Fertility Regulation. The association between vaginal bleeding patterns and reasons for discontinuation of contraceptive use. *Contraception* 38:207-225. 1988.

29. Belsey EM, Carlson N. The description of menstrual bleeding patterns: towards fewer measures. *Statistics in Medicine*, 10:267-284. 1991.

30. Treloar AE, *et. al.* Variation of the human menstrual cycle through reproductive life. *Int J Fertil* 12:77-126. 1967.

31. Chiazze L, *et. al.* The length and variability of the human menstrual cycle. *JAMA* 203:377-380. 1968.

32. Mayes DG, Harding CM. The statistical analysis of menstrual bleeding patterns. *Report of WHO Project 77084*, (unpublished).

33. Bernstein L, *et. al.* The effects of moderate physical activity on menstrual cycle patterns in adolescence: Implications for breast cancer prevention. *Br J Cancer* 55:681-685. 1987.

34. Ovellette MD, MacVicar MG, Harlan G. Relationship between percent body fat and menstrual patterns in athletes and non-athletes. *Nurs Res* 35:330-333. 1986.

35. Belsey EM et al. Determinants of menstrual bleeding patterns among women using natural and hormonal methods of contraception. I. The influence of individual characteristics. *Contraception* 38:243-257. 1988.

36. WHO Special Programme of Research, Development and Research Training in Human Reproduction; Task Force on Long-Acting Systemic Agents for the Regulation of Fertility. Multinational comparative clinical evaluation of two long-acting injectable contraceptive steroids: norethisterone oenanthate and medroxyprogesterone acetate. 2. Bleeding patterns and side effects. *Contraception* 17:395-406. 1978.

37. Sivin I. International experience with Norplant[R] and Norplant[R]-2 contraceptives. *Stud Fam Plann* 19:81-94. 1988.

38. WHO Special Programme of Research, Development and Research Training in Human Reproduction. Task Force on Oral Contraceptives. A randomized, double-blind study of six combined oral contraceptives. *Contraception* 25:231-241. 1982.

39. WHO Special Programme of Research, Development and Research Training in Human Reproduction. Task Force on Oral Contraceptives. A randomized, double-blind study of two combined and two progestogen-only oral contraceptives. *Contraception* 25:243-252. 1982.

40. Rodriguez G, Faundes-Latham A, Atkinson LE. An approach to the analysis of menstrual patterns in the critical evaluation of contraceptives. *Stud Fam Plann* 7:42-51. 1976.

41. Kruskal WH, Wallis WA. Use of ranks in one-criterion variance analysis. *J Am Stat Assoc* 47: 583-621. 1952.

VARIABILITY IN MENSTRUAL BLEEDING PATTERNS: COMPARING TREATED AND UNTREATED MENSTRUAL CYCLES

Siobán D. Harlow[1]

Dr. Belsey's paper begins with two fundamental concepts. First, interindividual differences in bleeding patterns associated with steroid contraception are neither random nor erratic. I make this concept explicit because, as is true for untreated menstrual cycles, it is only the conscious negation of the notion that menstrual bleeding patterns are erratic that leads us to define the nature of variability in these patterns and to search for determinants of this variability. Although our knowledge about the exogenous and endogenous determinants of variation in menstrual bleeding patterns remains limited, sufficient evidence exists to indicate that both inter- and intra-individual variation are associated with differences in host and environmental factors.

The second concept underlying Dr. Belsey's paper is that pharmacologic doses of exogenous hormones, such as steroid contraception, do not simply override the endogenous environment. Rather they interact with a dynamic endogenous system and the response of the endogenous environment is not necessarily the same in all women nor is it one of complete surrender. If we begin with these two concepts, we should, in fact, expect to see variability in bleeding characteristics associated with hormonal contraception consistent with the variability observable in untreated cycles.

Understanding the phenomenology of bleeding associated with the various steroidal contraceptives requires us to first consider what patterns and changes in patterns can be observed given different contraceptive formulations and routes of administration and this question has, in fact, been the focus of most previous research. However, as Dr. Belsey

S.D. HARLOW • Departmento de Epidemiología, Escuela de Salud Pública de México, Instituto Nacional de Salud Pública, Cuernavaca, Morelos, México

Steroid Contraceptives and Women's Response,
Edited by R. Snow and P. Hall, Plenum Press, New York, 1994

has suggested in her paper today, we must also ask how host and environmental factors may determine the inter- and intra-individual variations in these bleeding patterns. Other speakers in this workshop will later address this question from the perspective of pharmacokinetics.

I would like to suggest that we also need to be asking a third question: Are these patterns and responses consistent with expectations based on our understanding of the nature of variability in untreated menstrual cycles? Although our knowledge about variability and determinants of variability in untreated menstrual cycles also remains limited, the available information should nonetheless help us interpret bleeding data for contracepting women and help us to identify factors that are likely to influence their bleeding patterns.

Dr. Belsey has presented data that suggests it is possible to identify factors that partially explain the variability in bleeding patterns associated with specific types of contraceptives. The most influential of these factors is geographic region. European women, for example, tended to bleed longer and to have shorter bleed-free intervals than women in other parts of the world. As Dr. Belsey indicated, this pattern is consistent with data on regional differences in bleeding patterns for untreated menstrual cycles. Individual host characteristics, including menstrual and pregnancy history, were less informative; however, a high Quetelet index (eg. heavier women) was associated with longer bleed-free intervals.

Dr. Belsey concludes with the observation that separate from the effect of region, and possibly route of administration and weight, host characteristics had neither strong nor consistent effects in her data. Therefore, although the internal environment may not totally succumb to the effects of pharmacologic doses of estrogen, bleeding patterns may nonetheless be largely determined by contraceptive method.

The question that I would like to ask is what would we expect to see given her data and our knowledge of the nature of variability of menstrual bleeding patterns in the absence of hormonal contraception? I would agree with Dr. Belsey that contraceptive methods are a powerful determinant of bleeding patterns. Similarly in the untreated menstrual cycle, the major determinant of bleeding patterns is the dosing of hormones produced by the ovarian cycle, and variability in menstrual patterns is largely constrained by the nature of ovarian function. Thus whether we are examining treated or untreated segments, we can ask only how a risk factor may alter or perturb bleeding patterns within the parameters of variation imposed by the hormonal synthesis of the ovary or the contraceptive method. However, we have barely begun to consider the range of factors that may influence these bleeding patterns, and we still need further development of analytical methods that enable us to identify the effects of host or environmental characteristics.

If we first consider the nature of variability in untreated menstrual cycles, the distribution of menstrual segment lengths is such that the bulk of segments fall between about 20 and 40 days with a roughly symmetric distribution and a mode of 28 days, while the remainder create an extremely long right tail. The nature of this distribution has led us to think of menstrual bleeding segments as the sum of two parts: one part corresponding to the length of the ovarian cycle (e.g., the gonadotropin-dependent follicular and luteal phases) and the other part corresponding to the length of the waiting time from the end of one ovarian cycle to the successful initiation of the next ovarian cycle. We have found that thinking of a menstrual segment as the sum of these two parts has facilitated both our understanding of the nature of variability in menstrual bleeding patterns and our ability to identify risk factors for this variability[1,2].

In examining the nature of variability in menstrual bleeding patterns, we have found that women are heterogenous with respect to their probability of having a non-zero waiting time[1]. Also variability in the length of the ovarian segment can be characterized as a random effects process such that women tend to have a characteristic segment length, or set point, about which they vary, and which may be slightly longer or shorter than the population mean segment length. This structure of variation in segment length leads us to ask several questions about how a given risk factor may influence menstrual bleeding patterns:

1)   What factors are associated with determining differences in the set point within or between populations?, or

2)   Taking into account interindividual differences in the set point, what factors may perturb the length of the ovarian cycle (e.g. alter expected segment length)?, or

3)   What factors are associated with differences in the probability of having a non-zero waiting time (e.g. a long menstrual segment)?

Region appears to be a factor associated with differences in the set point for groups of women. It would be interesting to explore further what "region" is measuring (e.g., diet, weight, length of exposure to sunlight), as region is clearly a surrogate for a constellation of factors.

When we examined these questions in a population of 17-19 year-old freshman women[2] who were not using either hormonal contraception or the IUD, we were able to identify several factors which influenced the probability of having non-zero waiting time in a given segment. These factors included history of having long segments, starting college, experiencing life gain events, having a high level of perceived stress, dieting to lose weight, and level of physical activity. We also found that being at the upper end of the weight spectrum increased the probability of a long segment. Thus Dr. Belsey's finding, that heavier weight was associated with longer bleed-free intervals in contracepting segments, is consistent with what we might expect to see based on our finding of heavier weight being associated with a greater probability of a prolonged time between bleeds in a population-based sample of young women.

We were much less successful, on the other hand, in identifying determinants of variability in expected segment length given that the segment was shorter than 44 days. History of having long segments tended to lengthen and some stressors appeared to shorten expected length. As was true in Dr. Belsey's data, age of menarche did not appear to be influential in our data.

Our findings about the nature of variability in untreated menstrual cycles and the comparisons between our findings and Dr. Belsey's would lead me to make the following suggestions concerning methodological development and further substantive research on determinants of variability in bleeding patterns. First of all, we need to think about the fact that the range of variation in segment characteristics for most segments, regardless of the presence or absence of or type of contraception is actually relatively small. If the bulk of segment lengths fall in a 10 to 30 day range, and the bulk of bleed lengths fall in a three day range, then either very precise measurements will be required to see small shifts in mean length, or a risk factor would need to have a strong influence to exert a detectable effect. Perhaps it may be necessary to consider bleeding phenomena as something other than one continuous phenomenon, as it was necessary to evaluate the bipartite nature of menstrual segments in order to be able to observe the effects of risk factors. Alternatively our methods of analysis may need to focus on changes in the tails of the distributions as opposed to central tendency in order to identify risk factors which increase the probability of falling in the extremes of the distributions. We may also want to examine not only absolute differences in risk factor levels between individuals but also relative change in level of a risk factor within an individual. For instance, changing levels of physical activity or various stressors may provoke alterations in bleeding responses to a given contraceptive.

Furthermore, given our general lack of substantive knowledge, instead of focusing all of our efforts on investigating one factor such as weight, we should be asking what other host and environmental factors require evaluation before we assume that exogenous factors do not influence a woman's response to hormonal contraception. We may want to consider factors that mediate the endogenous hormonal environment or that interact with physiological processes which determine the endogenous hormonal environment. Other speakers will be considering this question when they discuss potential interactions of drug and hormonal metabolism, and the effect of diet on estrogen metabolism. However, by the same logic we may also want to consider the potential effect of chemicals widely dispensed in the environment including pesticides and other agricultural chemicals or industrial solvents and lead. If we borrow from our knowledge of untreated menstrual cycles, we should also consider investigating the effects of stress, or other constructs of the social environment, such as violence and intimidation, as well as physical activity.

Finally, I would suggest that our knowledge of determinants of variability in menstrual segments is generally inadequate. I would argue therefore for better integration of research on determinants of variability in bleeding characteristics in treated and untreated menstrual cycles. For if we do not understand either the endogenous hormonal environment or the systemic control and responses of our bodies to that environment which we manipulate with steroid contraception, it will be difficult to understand or to anticipate the bodies' responses to that pharmacologic intervention.

References

1. Harlow SD, Zeger SL. An application of longitudinal methods to the analysis of menstrual diary data. *Journal of Clinical Epidemiology* 44(10):1015-1025, 1991.
2. Harlow SD, Matanoski GM. The association between weight, physical activity and stress and variation in the length of the menstrual cycle. *American Journal of Epidemiology.* 133(1): 38-49, 1991.

CONTRACEPTIVE USE AND DISCONTINUATION: THE SOCIAL MEANING OF SIDE-EFFECTS

Carla Makhlouf Obermeyer[1]

In what follows I offer two sets of comments about contraceptive side-effects studies. The first is methodological, and concerns the appropriateness of generalizing from the type of follow-up study that is most commonly carried out. The second is conceptual, and presents a brief critique of the frameworks that are used to interpret the findings of contraceptive continuation studies. In discussing these points, I will draw mainly upon two disciplinary areas, epidemiology, and social science.

Before I begin with these general comments, let me make a few specific points about the study that Dr. Ahmed has reported on. It is based on a very rich data set that is analyzed in great detail with respect to the possibility of population variability. In the absence of baseline information about complaints in the population at large, the ratio of complaints from Norplant[R] compared to IUDs provides a very useful comparison of the rate and magnitude of side effects from the two methods. It is notable that, with one major exception, the study found no statistically significant differences in rates of complaints by country; the fact that Chinese women using the steel ring IUD suffer excess bleeding is congruent with the findings of other studies, and seems to confirm that this method is not the most appropriate to use in the Chinese population. In general, the good continuation rates and relatively low frequency of side-effects confirms the findings of other studies (Darney et al. 1990, Sivin 1988, Zimmerman 1990) that Norplant[R] has great potential for acceptability and widespread use.

C. OBERMEYER • Harvard School of Public Health, Department of Population and International Health, 665 Huntington Avenue, Boston, MA 02115

Steroid Contraceptives and Women's Response,
Edited by R. Snow and P. Hall, Plenum Press, New York, 1994

1.   The potential for selection bias: an epidemiological perspective

My first set of comments focuses on the overall design of the study, and the relevance of such a design for comparative studies of side-effects. The fact that Norplant[R] is a long-term contraceptive method has several implications for the methodology used in studying side-effects.   It means first that all of the women included in a study of continuation must have had strong motivation to use contraception continuously for an extended period.   It also implies-- a fact that is supported by data from other studies (Darney et al. 1990)-- that Norplant[R] users are "survivors" of other short-term contraceptive methods.   Consider the steps that are involved in studying the continuation rates and side-effects of acceptors, and then generalizing to the population at large. The population about which we want to make general statements is that of women in the reproductive years. We study "acceptors", and then compare the rate and type of complaints of those who continue using the method with those of the women who discontinue it.   But, as the following diagram illustrates (fig. 1), selection processes operate at several points as one goes from the population at large to the sample of "acceptors". and from the "acceptors" and the "continuers" and "discontinuers".   This means that the potential for selection bias is considerable, and that generalizations have to be made with caution.

In the case of Norplant[R], it may be that women who tend to have stronger side-effects have a lower probability of being enrolled in a study, of "surviving" the selection processes at work, and of having sufficient follow-up time to be included in the final comparisons.   If a large proportion of Norplant[R] users have in fact "withstood" the side-effects of other contraceptive methods, then the proportion of side-effects as measured in studies such as this one will be biased downward. If at the same time, the occurrence of stronger side-effects varies systematically by ethnic or national group, then this may lead to an underestimation of population differences. This possibility is actually mentioned by Dr. Ahmed, and I believe it needs to be given further consideration. One way to reduce the potential for selection bias is to use a life-table study design which would allow the inclusion of even those women with short follow-up times. But clearly, more needs to be done to collect baseline data on the rate and magnitude of complaints from women who, for a variety of reasons, are not included in the sample of post-marketing surveillance studies such as this one.

2.   Side-effects and complaints: a social science perspective

In the remainder of my commentary I wish to focus on the concept of side-effects from a broader perspective than the strictly biological, and consider the social and cultural context of contraception.   In most studies of side-effects, the assumption is that physical complaints are principally the assessment of a physical condition, and that they give a measure of satisfaction/dissatisfaction with a given product, in much the same way as other

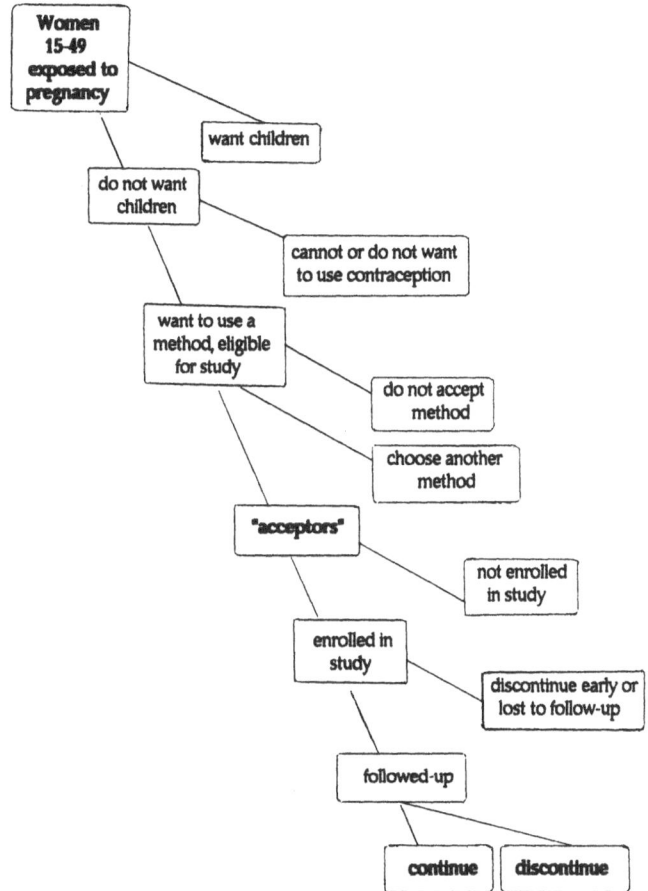

Figure 1. The potential for selection bias in studies of contraceptive side-effects.

consumer studies do. In a summary of the literature on the subject, a WHO report states that discontinuation due to method side-effects is the primary reason for abandoning a method or switching to another method (WHO 1988). I believe that such a view of complaints, while extremely valuable in clinical trials of a given method, becomes insufficient when one needs to understand the utilization rates of a larger population. It is important to broaden the meaning of complaints beyond what they seem to be saying about bodily experiences.

Reports of side-effects are related to notions about the body and its functioning, which are in turn influenced by the concepts and theories underlying various medical systems. There is a large body of anthropological literature on the variability of perceptions of the body among different cultures, and on the ways in which even scientific descriptions of the functioning of reproductive organs are influenced by cultural notions (Garrett 1984, Martin 1987). Particularly relevant here is an analysis of how Iranian women perceive contraceptives, which links these perceptions to ancient Galenic conceptions of body humors, and to classic Islamic medical theories concerning blood, the heart, natural temperament, and the reproductive system, (DelVecchio Good 1980). Another interesting study documents the way in which Jamaican women's perceptions of the effects of contraceptives reflect their images of the reproductive organs and their functions (McCormack, 1988).

These studies are of interest, not just because they illustrate the exotic belief systems that anthropologists sometimes encounter in the field, but because they portray women as actively synthesizing health information from a variety of sources. Whereas many clinical and marketing studies see women as passively responding to the physical stimuli caused by the various methods, micro-level participant observation studies show that women incorporate the information they receive about modern methods into their own understanding of the functioning of their bodies. This dynamic process of integration has a direct bearing on the acceptance and continuation of contraceptive methods.

More specifically, since the side effects reported in connection with most contraceptives, and with Norplant[R] in particular, center around disruptions of the menstrual cycle, it is important to carry out further research on the physiological, social and perceptual aspects of menstruation. This a topic that has received some attention both from social scientists and from those interested in family planning. In the mid-seventies, for instance, the WHO sponsored a collaborative study on menstruation in Egypt, India, Jamaica, Mexico, Pakistan, the Philippines, Korea, the UK, and Yugoslavia, which yielded some interesting insights into the variability in the experience of menstruation in different parts of the world (Snowden and Christian 1983). But such studies would have to be carried out prior to, or in combination with contraceptive continuation studies, if they are to constitute the sorely needed baseline for measuring and evaluating contraceptive side-effects. In general, social/anthropological research, if it could be conducted in conjunction with "mainstream" studies of contraceptive use, would add a new dimension to our understanding of the acceptance and continuation of contraception.

3.   Reported side-effects: symbolic aspects

One of the fundamental attributes of the anthropological approach is to consider social phenomena as multi-dimensional, that is relating to several levels of social experience. Anthropological analyses attempt to link the different aspects of a given observation, be they physical, economic, religious, or political. In this perspective, reports about side-effects do not merely describe facts about inconvenience, physical sensations, or pain, but also reveal ideas about power relations, gender roles, family norms, and societal values. Therefore, they have to be considered as multi-dimensional, symbolic statements. In certain social contexts, reports about side-effects may be a vehicle through which women express their general ambivalence towards contraception. It is clear that where a women's role is defined primarily as reproductive, and her status is enhanced principally through motherhood, then contraceptive methods will tend to be less acceptable, especially if they affect menstrual patterns. What is not often appreciated is that, faced with a conflict between the norms defining her status in her family, and those suggesting that family planning is a desirable choice, a woman may express her reasons for discontinuation in terms of side-effects, or attribute it to the opposition of family members (husband, mother-in-law), or ascribe it to the strength of religious prohibitions.

The extent to which such statements are as much symbolic as factual is illustrated by number of cases. In a study of oral contraceptives in Kuwait, a statistically significant association was found between holding beliefs that Islam is against family planning[1] and reporting side effects (Fakhr El Islam et al. 1988). Another study conducted in Bangladesh focused on 106 men whose wives had told the health workers that they were opposed to family planning. Interviews with these men revealed that of the alleged opponents, 26% were current users of family planning; an additional 50% were in favor of family planning though they themselves were not practicing it; 23% invoked religion, but only one fourth were able to cite specific religious statements about family planning to support their opposition (Bernhart and Moslehuddin 1990).

One of the many interesting implications of this study is that statements about family planning are to be understood at several levels, the factual being perhaps the least illuminating. This is an area of ambivalence, and of conflict among cultural norms. In hastening to categorize people as being 'for' or 'against' certain behaviors, many studies are missing one of the most important points: contraception affects the links between sex and reproduction, and therefore touches on two aspects of life that have, for a very long time, been considered as serious, even sacred. As the anthropology of religion has amply documented, man's relation to the sacred is characterized by ambivalence and sometimes fear, and is surrounded by many rules and sanctions. It is only in recent history that, for a greater segment of 'modern' populations, sex and reproduction have increasingly come to be dissociated from each other and from their implications. One could in fact think in terms of a continuum where at the one end would be the "ideal type" (in the Websterian sense) of a society where sex and reproduction are inextricably linked, and at the other end the type of society where they are completely separated, both technologically through contraception. and ideologically, though permissive social norms. Real societies would actually fall somewhere between these two extremes, closer to one end or the other.

The implications of this typology for research on contraception could be drawn at the individual level, by suggesting studies which would attempt to link discontinuation and improper use of contraception to individuals' deep attitudes towards the link between sex

---

[1]It should be noted here that most interpretations of Islamic doctrine by the different schools of law agree that family planning is acceptable (IPPF 1974, Omran 1980).

and reproduction. At the social level, one could envisage anthropological research in a given society on the norms and values that are directly or indirectly related to sex, reproduction and children. To do this would require a close examination of verbal statements and behavior of individuals, as well as the implicit and explicit statements that appear in the various media. On the basis of such an analysis, predictions could be made concerning the receptivity of the society, or of its various segments, towards contraception, and the degree to which individuals will clearly prefer to dissociate sex from its potential for reproduction.

Such an emphasis on ambivalence and rather than decisiveness would also call into question the prevalent models that are used to study contraceptive behavior. The typical

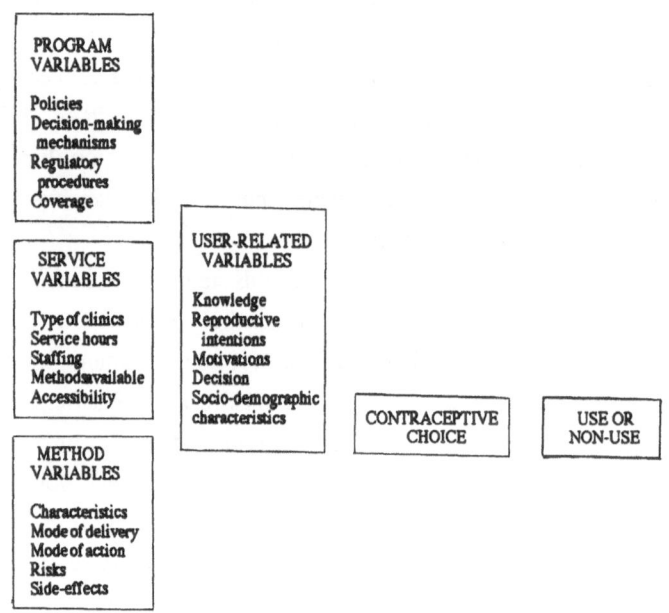

Figure 2. A Typical Model of Contraceptive Behavior.

framework used in most studies of acceptance and continuation can be illustrated in the following diagram, taken from a WHO report (1988).The outcome variable is dichotomised as use vs. non-use or continuation vs. discontinuation, and variables affecting it are categorized into program variables, service variables, user variables, and method variables. The assumption is of a 'user', rationally weighing convenience, cost, and control, and then making a decision based on a hierarchy of needs. Imperfect choices are seen as resulting from insufficient information or the inability of some users to be rational due to 'cultural' constraints. The problem with such models of contraceptive use is that the 'product', unlike detergents or toaster ovens, affects the meaning and outcome of a behavior that is tied to deep emotions, important rituals, and complex ideas related to life and death. A rational economic model of contraceptive behavior is therefore too simplistic. A more appropriate model would be built on indecisiveness rather than choice. The model would be longitudinal, and all women in the population could be included, and then followed-up to see whether they become 'acceptors', for how long when they stop contracepting and when they choose another method. Increasingly, the incorporation of data on method mixing and switching in studies of continuation is being advocated for all studies of contraceptive use (Bruce, 1989). It is now feasible, thanks to the development of hazard statistics, to build a model that can use the truncated data that result from the alternation among different methods. The major advantage is that the model would incorporate the shifting and constant reevaluation that characterizes much of social behavior.[2] An effort in this direction would thus do justice to the complexity of decision-making in this area of social life.

4. Conclusion

This paper has argued for a change in both the methodologies and conceptual frameworks used in studies of contraceptive continuation. It has called for a more rigorous definition of the sample studied and its relation to the population one is trying to generalize to. It has suggested the use of a more flexible methodology that would incorporate the truncated data that result from individuals shifting in and out of a given method. At the same time, it has advocated the use of the anthropological perspective to reach a broader understanding of the meaning of side-effects, and to consider reports of side-effects not merely as factual descriptions of bodily states, but also as statements about family structure, power relations, and the normative structure of society.

References

1. Bernhart M. and M. Mosleh Uddin. 1990. "Islam and family planning acceptance in Bangladesh". Studies in Family Planning 21(5):287-292.
2. Bruce J. 1989. "Fundamental elements of the quality of care: a simple framework". Population Council, Programs Division, Working Paper No. 1.
3. Darney P. et. al. 1990. "Acceptance and perceptions of Norplant[R] users in San Francisco, USA". Studies in Family Planning 21(3):152-160.
4. DelVecchio Good M. "Of blood and babies: the relationship of popular Islamic physiology to fertility". Social Science and Medicine 14B:147-156.
5. Fakher el Islam M.T. Malasi, and S. Abu-Dagga. 1988. "Oral contraceptives, socio-cultural beliefs, and psychiatric symptoms". Social Science and Medicine 27(9):941-945.
6. Garret K. 1984. "Women's reproductive rhythms: the nature/nurture conundrum". In: T. Duster and K. Garret (eds.). Cultural Perspectives on Biological Knowledge. Norwood, NJ: Ablex Publishing.

---

[2]This has been repeatedly documented in studies of health behavior which show individuals using both traditional and modern sources of care, and shifting from one to the other as the situation seems to dictate (see for instance Kleinman, 1980).

7. International Planned Parenthood Federation (IPPF). 1974. Islam and Family Planning. Report of an International Conference, Rabat, 1971. Beirut: Imprimerie Catholique.
8. Kleinman A. 1980. Patients and Healers in the Context of Culture. Berkeley: University of California Press.
9. Martin E. 1987. The Woman in the Body. Boston: Beacon Press.
10. Mernissi F. 1975. "Obstacles to Family Planning in Urban Morocco". Studies in Family Planning 6(12):418-425.
11. Omran AR. 1980. Population in the Arab World: Problems and Prospects. London: Croon Helm.
12. Sivin I. 1988. "International Experience with Norplant^R and Norplant^R-2 contraceptives". Studies in Family Planning 19(2):81-94.
13. Snowden R. and B. Christian (eds.). 1983. Patterns and Perceptions of Menstruation. Croom Helm and St. Martin's Press, in cooperation with WHO.
14. Young J. Medical Choice in a Mexican Village. New Brunswick: Rutgers University Press.
15. Zimmerman M. et. al. 1990. "Assessing the acceptability of Norplant^R implants in four countries: findings from focus group research". Studies in Family Planning 21(2):92-103.

POPULATION AND DELIVERY SYSTEMS: VARIABILITY IN PHARMACOKINETICS OF LONG-ACTING INJECTABLE CONTRACEPTIVES

Josue Garza-Flores, Sang Guo-wei and Peter E. Hall[1]

1. INTRODUCTION

Following the development and widespread use of oral hormonal contraceptives, it became evident that long-acting formulations of hormonal contraceptives could provide additional contraceptive options in some cultural settings where injectable or subdermal routes of administration are preferred. Despite a somewhat controversial history, long-acting contraceptives presently constitute an important option in family planning services[1]. Indeed, more than seven million women around the world are currently using long-acting injectable and implantable steroidal contraceptives, mostly in developing countries, and the number of users is increasing[2].

Worldwide research efforts during the past 25 years have focused on developing devices and formulations for the sustained release of contraceptive steroids. During the course of numerous pharmacokinetic and pharmacodynamic investigations and other clinical studies it became evident that differences among human populations exist[3-5].

A number of factors which can account for such ethnic differences have been advocated, both environmental and genetic ones. It is well known that alterations in diet and nutritional status may influence responses to drug therapy. Ethnic differences in the metabolism of sex steroid hormones and in the prevalence of hormone-related disease, including tumours arising in hormone-response tissues such as the breast and endometrium have been documented[6-7].

J. GARZA-FLORES • Instituto Nacional de la Nutrición S.Z., Departamento de Biologiá de la Reproducción, Tlalpan Vasco de Quiroga, 14000 México, D.F.

Nutrition-induced changes in steroid biotransformations have the potential for substantially modifying both the hormonal action of steroids and their biologically active metabolites. Differences in the distribution of fat together with adipocyte metabolism may be a greater influence on the pharmacokinetics of steroids which are continuously released from depot sites, such as those created by deep intramuscular injections or by delivery systems such as implants or vaginal rings[8,10]. The purpose of this review is to describe the population differences found during the course of multicentred assessment of the long-acting injectable contraceptive formulations, depot medroxyprogesterone acetate (DMPA) and norethisterone enanthate (NET-EN).

2. PHARMACOKINETIC STUDIES IN OBESE AND THIN WOMEN

Results from clinical trials have shown that the progestogens medroxyprogesterone acetate (MPA) and norethisterone enanthate (NET-EN), both of which have been in clinical practice for more than 20 years, are highly effective contraceptive agents[1,11]. Pregnancies due to method failure have been consistently low (cumulative life-table rates at 12 months of 0-0.1 percent) with the use of 150 mg DMPA administered every 90 days. The pregnancy rates reported with NET-EN use varied according to the injection schedule used. NET-EN given at 60-day intervals gave rise to cumulative life-table rates of 0.4% at 12 months and 0.4% at 24 months, while when given at 60 days for six months and subsequently at 84-day intervals they were 0.6% at 12 months and 1.4% at 24 months.

In the mid-1970s, a large randomized multicentred comparative trial was organized by the World Health Organization (WHO)[5] to assess the use-effectiveness of 200 mg of NET-EN and 150 mg of DMPA given every 12 weeks $\pm$ 5 days. The study involved 1,678 women and although it was supposed to follow subjects for up to two years, it was stopped after one year because a total of 28 pregnancies were reported (24 for NET-EN and four for DMPA). The cumulative life-table pregnancy rate at one year was 3.6 percent for NET-EN and 0.7 percent for DMPA. Two of the ten centres (Bangkok and Chandigarh) contributed 54 percent of all pregnancies in the NET-EN group, and 75 percent of all conceptions with NET-EN were estimated to have occurred during the first injection period, mostly in the last four weeks of this interval. It was found that body weight of NET-EN users who became pregnant was significantly lower than those who did not (mean of 7.02 kg less), whereas there was no such difference among subjects treated with DMPA. From these results two issues became apparent: a) that the injection schedule for NET-EN had to be modified (to every two months); and b) that body weight might influence contraceptive effectiveness particularly with regard to long-acting injectable compounds. Accordingly, pharmacokinetic studies of both DMPA and NET-EN in obese and thin women of reproductive age of different populations were undertaken.

Table I. Pharmacokinetic parameters in three studies on DMPA (150 mg) in Thai and Mexican women.

Study	No.	Quetelet Index	Cmax (nmol/l)	Tmax (days)	T1/2 (days)	AUC(nmol/l/24h)	Tc <0.259 nmol/l (days)
1. - Thai thin	10	18.0 ± 1.8	16.5 ± 6.1	7.0 ± 3.2	27.4 ± 22.8	606 ± 307	160 ± 73
- Thai obese	10	32.2 ± 2.8	18.2 ± 15.7	8.6 ± 4.9	30.2 ± 18.1	582 ± 326	145 ± 42
- Mexican thin	5	18.6 ± 0.6	2.3 ± 0.9	22.4 ± 14.5	56.4 ± 42.8	96 ± 22	171 ± 95
- Mexican obese	5	31.1 ± 3.0	2.4 ± 0.8	8.8 ± 3.0	112.2 ± 117.8	93 ± 25	305 ± 274
2. - Thai	5	23.1 ± 2.4	21.5 ± 8.3	-	24.1 ± 23.6	558 ± 79	92 ± 44
- Mexican	4	22.9 ± 2.1	3.4 ± 5.2	-	62.5 ± 23.7	259 ± 152	187 ± 85
3. - Thai	14	22.1 ± 2.1	25.9 ± 16.3	8.0 ± 3.7	24.3 ± 18.1	606 ± 352	143 ± 78

The results from these studies[4,12-14] are summarized in Table I. They showed that, although no significant differences were observed between obese and thin women in many of the pharmacokinetic parameters, there was a tendency to more rapid absorption of the contraceptive steroid in thin as compared with obese women. However, when the data from the two participating centers in Bangkok and Mexico City were compared, major differences in pharmacokinetic parameters were observed. Peak values of MPA (Cmax) were, in most subjects, almost an order of magnitude greater in Thai women than in Mexican women, and the area under the curve was some six-fold greater. It should be noted that the same formulations of DMPA were used in these studies and that all hormone assays were undertaken by the same laboratory.

3. DOSE REDUCTION STUDIES

In order to assess the effects of different doses of DMPA on ovarian function, a study on the pharmacokinetic and pharmacodynamic effects of DMPA at doses of 25, 50, 100 and 150 mg were conducted in Bangkok and Mexico City. The results in Table II showed higher peak serum levels of MPA (Cmax), a more rapid disappearance of MPA from the blood and a more rapid return of ovulation at all dose levels in Thai women compared with Mexican women[5,16]. Table III summarizes the period of suppression of ovulation in the two populations. DMPA suppressed ovulation for longer periods in Mexican women at all dose levels.

This pharmacodynamic difference was reflected in the pharmacokinetic profiles. Although Quetelet Indices were similar in both populations, maximum MPA concentrations (Cmax) in serum were significantly higher in Thai women than in those observed in Mexican women at all doses. Thai women consistently gave rise to a greater area under the serum concentration curve (AUC), a shorter elimination half-life (T1/2), and reached serum concentrations below the limit of detection (Tc<0.259 nmol/l) more rapidly.

Since serum levels of MPA were not significantly different between 100 and 150 mg doses of DMPA, it was postulated that a dose of 100 mg of DMPA would be more than adequate to provide full three-month contraceptive coverage, particularly in Mexican women. To answer this question, a multicentred study comparing 100 mg or 150 mg DMPA was undertaken[17]. The results showed little difference in efficacy and side-effects between the two treatment groups, despite which there was hesitancy in recommending a reduction of the dose of DMPA globally because of pregnancies (cumulative pregnancy rate of 0.4%) occurring in two women on the 100 mg dose. The women had Quetelet Indices of 16.7 and 21.3. Improved formulation of DMPA may assist in allowing dose reduction, maybe as much as half, but until such formulations are available the increased risk of pregnancy, albeit small, might adversely affect acceptance of such a highly effective method of contraception.

Table II. Pharmacokinetic parameters of different doses of DMPA in Thai and Mexican women.

	THAI WOMEN				MEXICAN WOMEN			
MPA Dose (mg)	25	50	100	150	25	50	100	150
No. of subjects	5	5	5	5	3	4	4	4
Quetelet Index	21.8 ± 1.7	24.8 ± 3.7	22.1 ± 3.3	23.1 ± 2.4	24.8 ± 4.0	23.7 ± 1.5	25.4 ± 3.5	22.9 ± 2.1
Cmax (nmol/l)	5.5 ± 3.7	10.1 ± 4.6	16.5 ± 5.6	21.4 ± 8.2	1.51 ± 0.2	2.8 ± 0.7	2.8 ± 0.9	3.4 ± 5.3
T½ (days)	14.4 ± 7.3	11.8 ± 3.0	18.4 ± 8.7	24.1 ± 23.6	32.6 ± 7.6	42.0 ± 11.6	48.4 ± 16.1	62.5 ± 23.7
AUC (nmol/l/24h)	142 ± 16	197 ± 57	447 ± 73	566 ± 79	137 ± 64	142 ± 42	236 ± 76	258 ± 152
Tc < 0.259 nmol/l (days)	44.6 ± 11.3	52.6 ± 10.8	80.6 ± 24.3	92.0 ± 44.2	112.1 ± 27.9	133.7 ± 27.8	184.8 ± 42.2	186.7 ± 84.8

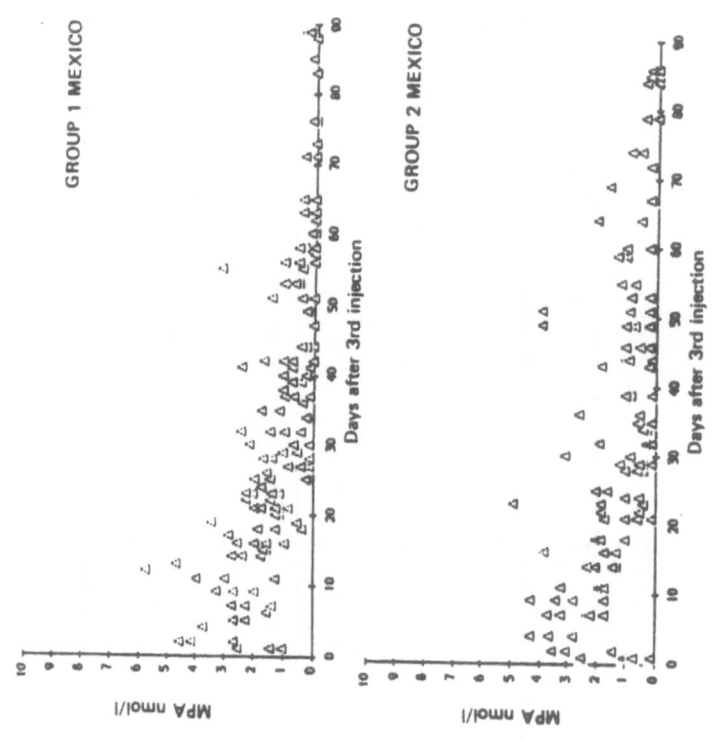

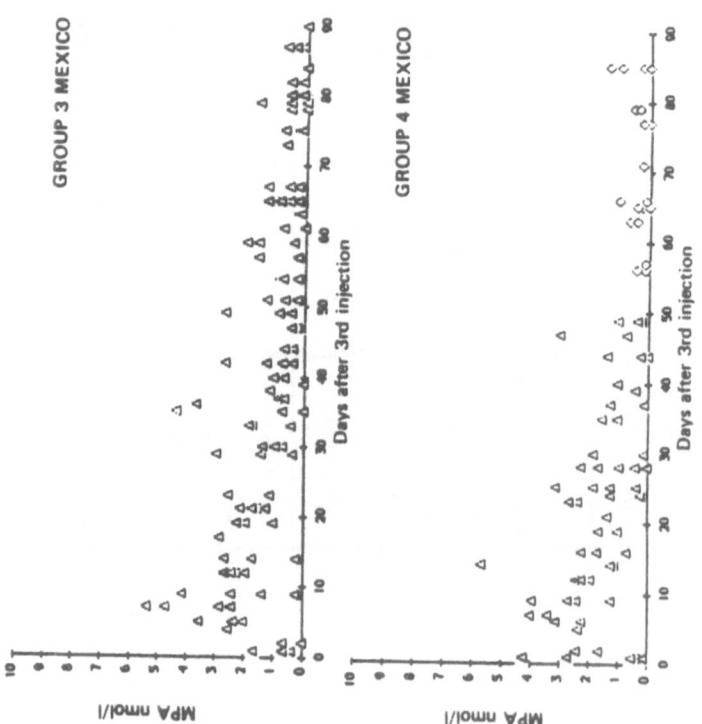

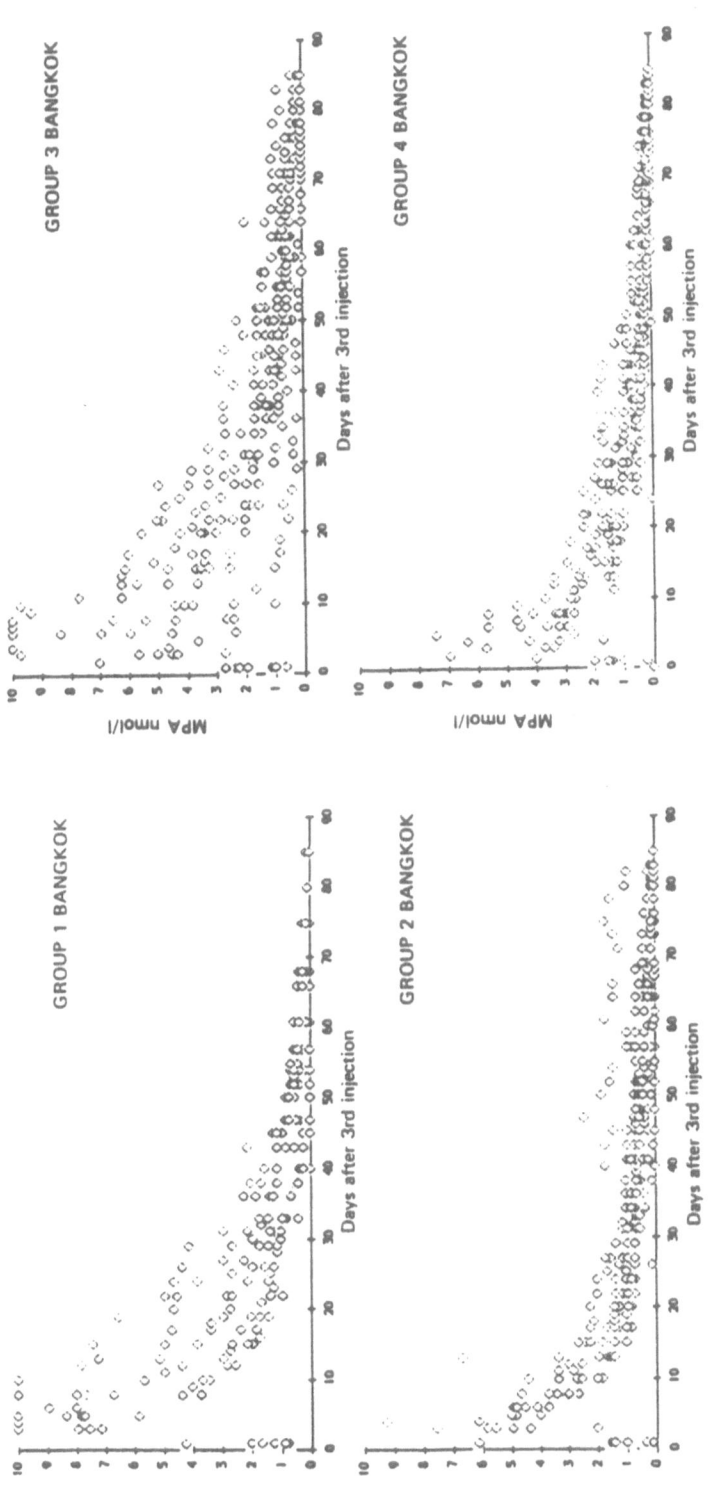

Figure 1. Serum levels of medroxyprogesterone acetate (nmol/l) in Bangkok women (A) and in Mexico City women (B) during the third treatment interval with HRP 112 Full (DMPA 25mg plus Estradiol Cypionate 5mg), HRP Half (DMPA 12.5mg plus Estradiol Cypionate 2.5 mg), DMPA 25mg or DMPA 12.5. Modified from reference 19.

4. ONCE-A-MONTH COMBINED INJECTABLE FORMULATIONS

WHO's Task Force on Long-Acting Systemic Agents for Fertility Regulation has been developing new once-a-month injectable preparations, combinations of a progestogen andan estrogen, which give rise to a significantly lower incidence of menstrual irregularities than seen with progestogen-only preparations[3]. Thus various combinations of DMPA with estradiol cypionate (E2-Cyp) and NET-EN with estradiol valerate (E2-Val) at different doses have been studied[18,19]. The study on DMPA-containing preparations was undertaken in Bangkok and Mexico City. Figure I shows the actual MPA levels measured in each serum sample. With each of the four formulations, the Cmax levels in Mexican women were 30-60 percent of those observed in Thai women, showing exactly the same relationship as those studies reported in Table I, whether the DMPA is given alone or in combination with an estrogen.

5. PARTICLE SIZE AND DURATION OF CONTRACEPTIVE EFFECT

Some years ago, it was reported in Chiang Mai that there was an increase in method failure with DMPA when a locally produced formulation was utilized by the Thai National Family Planning Programme instead of Depo-provera[20]. It was found that the batches locally manufactured, while being to pharmacopeia specifications, had a smaller particle size distribution of the micronized crystals than the previously used batches of Depo-provera. It was postulated that the smaller particles were absorbed more rapidly than larger ones causing MPA to appear more rapidly in the circulation. This, coupled with the more rapid elimination of DMPA in Thai women, could possibly give rise to a situation whereby small particle formulations provide less than 90-day contraceptive protection in this population.

A multicentred pharmacokinetic study was, therefore, designed to assess the influence of particle size on the pharmacokinetics of DMPA in Thai and Mexican populations. The preliminary data analysis as presented in Table IV again confirmed the differences previously observed in these populations. All 15 subjects in Mexico City had a Cmax of less than 10 nmol/l, while in Chiang Mai seven of the 20 women had Cmax values between 12.7 and 38.0 nmol/l. Of these seven subjects, five received the formulation of particle size range 0.1-0.5 um (Fisher's exact test: $p < 0.05$). Because of the almost inevitable problem of the small number of subjects participating in pharmacokinetic studies of this type, it is impossible to draw firm conclusions, however, the data suggest that the particle size of DMPA formulation is important for its duration of action and, therefore, contraceptive protection. Additional data from a further 20 subjects in Bangkok is presently awaited.

6. NORETHISTERONE ENANTHATE

The other widely used injectable progestogen preparation, norethisterone enanthate (NET-

Table III. Return to Ovulation in days after injection of various doses of DMPA in Mexican and Thai women.

Dose (mg)	Mexican	Thai	Statistical Significance
25	127 ± 29.9	74.6 ± 23.8	P < 0.05
50	165 ± 23.9	94.3 ± 27.2	P < 0.01
100	154 ± 47.7	131 ± 19.5	P < 0.01
150	180 ± 0	136 ± 24.5	P < 0.01

Each group contained five women.

EN), is formulated not as a microcrystalline suspension like DMPA but as an oil solution of benzyl benzoate and castor oil[21]. An intramuscular injection of 200 mg of NET-EN was shown to result in peak serum levels of NET within the first week followed by a gradual decline. As with compounds, considerable individual variation has been reported[22]. However, as can be seen from the limited number of studies reported in Table V, there is little obvious difference between populations. However, results on Chinese women who received 200 mg NET-EN were compared with those previously obtained by the same investigator in British women[23]. There was no difference between Chinese and British women in the absorption after intramuscular injection, but the elimination rate was significantly lower and the bioavailability higher in Chinese women than in British women[24].

Recent studies in Mexican women on the pharmacokinetics and pharmacodynamics of NET-EN have demonstrated that no major differences are observed with those obtained in European women except that there is a tendency for ovulation to return later in Mexican women. An additional study is ongoing in Indonesia.

7. DISCUSSION

All the studies reported above show one clear-cut result: when exposed to a micro-crystalline suspension of a steroid such as DMPA, Thai women absorb it more rapidly, giving rise to higher serum peak concentration values, and eliminate it more rapidly, giving a faster disappearance from the blood stream than Mexican women. This finding becomes virtually irrefutable since it is replicated in study after study whatever the dose injected intramuscularly and despite the inevitable wide subject variability and the inevitable small numbers of subjects assessed in such studies.

It is in fact remarkable to observe pharmacokinetic profiles of MPA in Mexican women following injection with 150 mg of Depo-provera. Cmax values are invariably between 2 and 7 nmol/l (see Table I, Studies 1 and 2, and Table V, formulation 1) and measurable concentrations of MPA are invariably observed six months after the injection. However, even low doses of DMPA still give effective MPA blood levels for long periods of time (see Table V and Figure I) to the point that a Mexican family planning program manager might speculate whether 150 mg of DMPA is an appropriate 90-day injectable contraceptive in Mexico.

Contrast this with the Thais, where there is a far greater subject variability with many subjects giving blood levels an order of magnitude greater than their Mexican counterparts with a range of Cmax values of 4 to 40 nmol/l with a significant proportion greater than 20 nmol/l (see Table I and Table V, formulation 1). The Thais have rapid elimination half-lives and often

Table IV. Preliminary analysis of DMPA (150mg) of different particle sizes administered to women from Thailand and Mexico.

	Quetelet Index (wt/h²)	Cmax (nmol/l)	Tmax (days)	K_e	T½ (days)	C90 days nmol/l	AUC (0-90 days) nmol/l/24h
Chiang Mai (n = 10 per group)							
Formulation 1 (0.5 μm)	19.6 ± 2.0	17.1 ± 12.0	3.7 ± 2.0	0.0139 ± 0.0106	125.7 ± 169.1	4.0 ± 1.8	578 ± 198
Formulation 2 (10-25 μm)	20.7 ± 1.8	8.3 ± 4.9	4.3 ± 3.2	0.0083 ± 0.0065	126.0 ± 98.0	3.7 ± 1.6	418 ± 199
Mexico City (n = 8 per group)							
Formulation 1 (0.5 μm)	22.3 ± 2.7	5.6 ± 1.6	4.6 ± 3.3	0.0058 ± 0.0027	130.3 ± 57.9	1.9 ± 0.8	212 ± 46
Formulation 2 (10-25 μm)	27.5 ± 8.8	4.4 ± 0.9	6.7 ± 3.5	0.0092 ± 0.0127	197.5 ± 163.4	2.5 ± 1.1	226 ± 33

Table V. Pharmacokinetic/Pharmacodynamic parameters after NET-EN (200 mg) in different populations.

Population	Cmax (nmol/l)	Tmax (days)	Return to ovulation (days after injection)	Reference
USA (n = 5)	44.9 (+/- 18.09)	7.1 (4-15)	98 (45-120)	[26]
UK (n = 8)	37.2 (+/- 8.3) (13.7-87.1)	4.3 (2-8)	–	[24]
Sweden (n = 14)	34.2	6.0	–	[27]
(n = 4)	47.6 (28.5-65.7)	3.5	85	[25]
India (n = 4)	50.9 (16.7-83.7)	5.0	99 (110-130)	[25]
Mexico	55.74 (19.6-116.95)	5.36 (2-10)	107 (86-126)	[28]

undetectable blood levels at 90 days, the time of next injection. These differences are equally pronounced when one looks at the area under the curve (AUC) between the day of injection and day 90.

Thai family planning program managers have different issues to consider than their Mexican opposite numbers, such as how can they be sure that 150 mg is always an effective dose over a 90-day period? The Thai woman is likely to be sensitive to formulation issues such as the particle size of the preparation or the effect of drug or nutritional interactionswhich might increase the rate of metabolism of the drug. Hence the hesitancy to recommend dose reduction following the major Phase III clinical trials of 100 mg and 150 mg DMPA on a global basis.

Studies are ongoing to assess whether these observations are characteristic of all microcrystalline formulations. Pharmacokinetic profiles of levonorgestrel butanoate have been generated in Mexico, London and Szeged and are now being assessed in Bangkok and Jakarta. Certainly there seems to be less exaggerated differences in the pharmacokinetics of the NET-EN dissolved in an oily solution, although Chinese women do eliminate it faster than British women, and Mexican women do return to ovulation later than other population groups (See Table IV). Even if population differences are more pronounced with microcrystalline suspensions, what are the underlying reasons behind them? An initial idea was that it was due to body size. This, however, appears to be clearly refuted by the data summarized in Table I, Study 1 where the large differences in Quetelet Index in each group in Mexico and Thailand are completely submerged by the overall Thai/Mexican differences. That is not to say that differences in fat distribution and/or adipocyte metabolism may influence steroidal pharmacokinetics in different populations.

It may well be that differences in the distribution of fat in individuals of similar body mass which coupled with the fact that fat in different anatomical sites may exhibit different metabolic characteristics[8-10]. It is quite feasible to consider that fat acts as a secondary depot for steroids after their initial release from the primary depot in muscle. Hence, ethnic differences in fat distribution and the behavior of fat cells in steroid storage could play a major part in explaining the ethnic differences in the pharmacokinetics of steroids described.

There are, however, many other factors which could influence steroid pharmacokinetics. For example, differences in plasma-binding of drugs have been described between caucasians and Chinese subjects[6]; dietary factors and other endogenous substances can affect liver metabolism[7]; physical activity can affect the dispersion of a steroid depot in muscle tissue; and so on. However, the authors consider that there would be great value in undertaking studies on fat distribution and metabolism and their influence on steroid pharmacokinetics in Thai and Mexican women. Whether large enough group sizes can be recruited to make such studies statistically valid remains to be seen, but maybe it is possible given the enormous difference between Thais and their Mexican sisters.

Acknowledgements

Thanks are due to the more than 200 women, mainly Mexican and Thai, who volunteered to have innumerable venepunctures to allow pharmacokinetic profiles to be analyzed and speculated over; the investigators, doctors and nurses who looked after them; and WHO's Special Programme for Research in Human Development for funding the studies.

References

1.   Anderson KE, Conney AH, Kappas A. Nutrition as an environmental influence on chemical metabolism in man. *Ethnic Differences in Reactions to Drugs and Xenobiotics.* pp. 39-45. 1986.

2.   Bardin CW, *et. al.* Norplant^R contraceptive implants. In *Fertility Regulation Today and Tomorrow.* Diczfalusy E, Bygdeman M (eds.). Raven Press, New York. pp. 119-141. 1987.

3.   Bassol S, *et. al.* Ovarian function following a single administration of depot-medroxyprogesterone acetate (DMPA) at different doses. *Fertil Steril* 42:216-222. 1984.

4.   Cravioto MC, *et. al.* Depot-medroxyprogesterone acetate and norethisterone enantate in obese and thin women (Unpublished data).

5.   Facts about injectable contraceptives: Memorandum from a WHO Meeting. *Bull WHO* 60:199-210. 1982.

6.   Fotherby K, *et. al.* Plasma levels of norethisterone after single and multiple injections of norethisterone oenanthate. *Contraception* 60:1-6. 1978.

7.   Fotherby K, *et. al.* A preliminary pharmacokinetic and pharmacodynamic evaluation of Depo-medroxyprogesterone acetate and norethisterone enanthate. *Fertil Steril* 34:131-139. 1980.

8.   Fotherby K, Koetsawang S. Metabolism of injectable formulations of contraceptive steroids in obese and thin women. *Contraception* 26(1):51-58. 1982.

9.   Fotherby K. Factors affecting the duration of action of the injectable contraceptive norethisterone enantate. *Contraception* 2:249-257. 1981.

10.   Fotherby K, Koetsawang S, Mathrubutham M. Pharmacokinetic study of different doses of Depo-provera. *Contraception* 22:527-536. 1980.

11.   Gabelnick HL. Biodegradable implants: Alternative approaches. In *Advances in Human Fertility and Reproductive Endocrinology.* Mishell Jr., DR (ed.). Raven Press, New York. 1983.

12.   Garza-Flores J, *et. al.* A pharmacokinetic and pharmacodynamic study of norethisterone enanthate in Mexican women. *Contraception* 1990 (in press).

13.   Goebelsman V, *et. al.* Serum norethisterone (NET) concentrations following intramuscular NET enanthate injection: Effect upon serum LH, FSH, estradiol and progesterone. *Contraception* 19: 283-313. 1979.

14.   Hall PE, d'Arcangues C. Long-acting methods of fertility regulation, World Health Organization, Special Programme of Research, Development and Research Training in Human Reproduction. *Research in Human Reproduction Biennial Report 1986-1987.* pp. 129-150. 1988.

15.   Kissebah AH, *et. al.* Relations of body fat distribution to metabolic complications of obesity. *Journal of Clinical Endocrinology and Metabolism* 54(2):254-192. 1982. (sic)

16.   Krotkiewski M, *et. al.* Impact of obesity on metabolism in men and women: Importance of regional adipose tissue distribution. *J Clin Invest* 72:1150-1162. 1983.

17.   McDaniel EB, Gray RH, Pardthaisong T. Method failure pregnancy rates with Depo-provera and a local substitute. *Lancet* i:1293. 1980.

18.   Rowe PJ. Steroid-releasing vaginal rings. A review. In *Fertility and Sterility.* Harrison RF, Bonnar J, Thompson W (eds.). MTP Press Ltd, Lancaster, UK. pp. 301-309. 1984.

19.   Sang GW. Personal communication (1987).

20.   Sang GW, *et. al.* Pharmacokinetics of norethisterone oenanthate in humans. *Contraception* 24:15-27. 1981.

21.   Toppozada M. Norethisterone (Norethindrone) enanthate clinical studies. In *Long-Acting Contraceptive Delivery Systems.* Zatuchni GI, Goldsmith A, Shelton JD, Sciarra JJ (eds.). Harper and Row: Philadelphia. pp. 502-514. 1984.

22.   World Health Organization, Task Force on Long-Acting Systemic Agents for Fertility Regulation: A multinational comparative clinical trial of long-acting injectable contraceptives; norethisterone enantate given in two dosage regimens and depot-medroxyprogesterone acetate - final report. *Contraception* 28:1-20. 1983.

23.   World Health Organization, Task Force on Long-Acting Systemic Agents for Fertility Regulation: A multicentred phase III comparative trial of 150 mg and 100 mg of depot-medroxyprogesterone acetate given every three months: Efficacy and side-effects. *Contraception* 34:223-235. 1986.

24.   World Health Organization, Task Force on Long-Acting Systemic Agents for Fertility Regulation, Special Programme of Research, Development and Research Training in Human Reproduction: A multicentred pharmacokinetic/pharmacodynamic study of once-a-month injectable contraceptives. I. Different doses of HRP 112 and of Depo-provera. *Contraception* 36:441-457. 1987.

25.   World Health Organization, Task Force on Long-Acting Systemic Agents for Fertility Regulation, Special Programme of Research, Development and Research Training in Human Reproduction: A multicentred pharmacokinetic/pharmacodynamic study of once-a-month injectable contraceptives. II. Different doses of HRP 102 and of norethisterone enanthate. *Contraception* 1990 (in press).

26.   World Health Organization, Special Programme of Research, Development and Research Training in Human Reproduction. *Research in Human Reproduction, Biennial Report 1988-1989.* pp. 17-40. 1990.

27.   World Health Organization, Special Programme of Research, Development, and Research Training in Human Reproduction; Task Force on Long-Acting Systemic Agents for the Regulation of Fertility: Multinational comparative clinical evaluation of two long-acting injectable contraceptive steroids: norethisterone oenanthate and medroxyprogesterone acetate. *Contraception* 15(5):513-533. 1977.

28.   Zalanyi S, Sandgren BM, Johannison E. Pharmacokinetics, pharmacodynamic and endometrial effects of a single dose of 200 mg norethisterone enanthate. *Contraception* 30:225-237. 1984.

29.   Zhou HH, Adedoyin A, Wilkinson GR. Differences in plasma binding of drugs between Caucasians and Chinese subjects. *Pharmacokinetics and Drug Disposition* 48:10-17. 1990.

PHARMACOLOGY OF THE ETHYNYL ESTROGENS IN VARIOUS COUNTRIES

Joseph W. Goldzieher[1]

By 1985, there were already a dozen or so pharmacokinetic studies of ethynyl estradiol in different countries. The data were noteworthy for the great range of the various pharmacokinetic parameters[1]:

$t_{1/2}\alpha$:	0.5	-	2.4hr
$t_{1/2}\beta$:	13.1	-	27.0hr
$t_{1/2}(Ka)$	0.2	-	0.4hr
bioavail.	0.38	-	0.48
K_{12}	0.18	-	0.25/hr
K_{21}	0.10	-	0.25/hr
t_{max}	1	-	2hr

The problem becomes one of separating out variance due to technological problems from inter-individual and inter-group differences, the latter being the ultimate aim of this review.

The technique of radioimmunoassay of ethynyl estradiol has undergone major improvements in the last decade. Of particular importance was the issue of specificity, since endogenous estradiol was a potential cross-reactant, and even more importantly the relatively huge amounts of accompanying 19-nor progestins were a problem even at levels of cross-reactivity which are acceptable for most RIA procedures. The latest techniques[2] obviate this problem and also deal adequately with the problem of nonspecific cross-

J. GOLDZIEHER • Visiting Professor, Texas Technological University, Amarillo, Texas. Address correspondence to: 626 Metropolitan Professional Building, 1303 North McCullogh, San Antonio, Texas, 78212.

Steroid Contraceptives and Women's Response,
Edited by R. Snow and P. Hall, Plenum Press, New York, 1994

reactivity. On the other hand the sensitivity of these assays, even in the best of hands, is far from satisfactory, given the progressively diminishing dosage that is being employed clinically. Plasma levels at 8, 12 and 24 hours after a single dose are at the limit of detectability and these values are highly important for the study of cumulative levels as well as in clinical considerations of efficacy. While it is reasonable to assume that 2, 3, or 4 times the clinical dose can be used for pharmacokinetic studies, it would be preferable to carry out studies reflecting actual clinical usage.

Space does not permit a review of the usual caveats to pharmacokinetic studies - particularly the possibility of differences in tablet dissolution and absorbability from one brand of oral contraceptive to another, the influence of stomach contents on an agent which is absorbed very rapidly from the stomach, and the problem of enterohepatic recycling, which can be affected by food intake at any time during the observation period.

Sampling time and especially sampling frequency have a large impact on parameters such as α, β and Ka; this issue has been discussed in some detail by Newburger and Goldzieher[1]. Adequate sampling frequency is also important in an effort to minimize the error introduced by assay insensitivity in the tail of the plasma level curve.

The pharmacokinetics of ethynyl estradiol has in most cases been described as a 2-compartment model after both IV and oral administration. One report found the data to be consistent with a 3-compartment model[3]. Humpel et al. have pointed out that this discrepancy could arise from the very short half-life of the second distribution phase. The average $t_{1/2}$ reported for this phase after IV administration was approximately 9 minutes. In one hour this phase will have gone through seven half-lives and will have essentially disappeared at the end of this time. If only one data point were obtained during this time, the second distribution phase could not be identified and would merge with the other distribution phase. In a 3-compartment model there are 7 parameters to be identified. A rule of thumb is that there should be 3 data points for each parameter, or in this case 21 data points. It may be possible to use fewer points if they are taken at strategic times among the different phases. However, in most studies the tail of the curve, where assay error is the largest, has been inadequately sampled.

Another issue which complicates ethynyl estradiol pharmacokinetics is the role of the substantial concentrations of mono and di-sulfates in the circulation. Our studies[4] suggest that the re-conversion of sulfates to free ethynyl estradiol does not make a significant contribution to plasma levels and therefore does not constitute a "reservoir"; Dr. Fotherby and I appear not to agree on this point.

The number of subjects in each clinical trial becomes a matter of great importance, considering what we have learned in recent studies of inter- and intra-individual variation. We have published a comparative pharmacokinetic study of three bioequivalent ethynyl estradiol/norethindrone and three mestranol/norethindrone formulations[5]. There were 24-27 subjects in each group, and they were tested with a single dose of a randomly-assigned formulation in the follicular phase of three consecutive menstrual cycles.

Given the observed degree of inter-individual variation (which will be discussed below), the AUC of mestranol, 50 mcg., was the same as from 35mcg. ethynyl estradiol given as such. This raises some interesting questions as to the validity of discussions of "high-dose" versus "low-dose" pills where the identity of the estrogen was not taken into consideration. In the present context, we must point out that the efficiency of conversion of mestranol into plasma levels of ethynyl estradiol in this group of subjects does not necessarily reflect what might happen in populations which differ ethnically, nutritionally, or in some other characteristic.

The data on AUC_{0-24} yielded a value of 1036 ± 483 pg.hr/ml, a C_{max} of 174 ± 67 pg/ml, a T_{max} of 1.3 ± 0.5 and a K_{el} of 0.90 ± 0.09. Individual values for AUC ranged from 234 to 2498 pg.hr/ml, with a coefficient of variation (CV) of 47%. The minimum and maximum C_{max} values ranged from 55 to 311 pg/ml with a CV of 39%. These very large variances are characteristic of drugs with a first-pass effect. Their importance, however, is that they add a considerable degree of uncertainty to comparisons based on small numbers of subjects, as for example in our studies in Sri Lanka, Nigeria and several USA localities. Were it not for qualitative differences in urinary metabolite patterns in various localities, as summarized below, the geographic differences in pharmacokinetics which have been reported by many investigators would rest on somewhat uncertain ground.

The study just described[5] also permitted an examination of a less well-studied aspect - that of intra-individual variation from one cycle to the next. It is well known that "normal" menstrual cycles vary considerably from month to month in endocrine values such as estrogen or progesterone levels without substantial differences in cycle duration and bleeding characteristics. Less is known about oral contraceptive steroid levels from cycle to cycle.

The within-subject coefficient of variation of AUC in this study was 41%. Taking the average of the three, AUCs stretched from -7% to +92% of the mean of the individual. Both the inter-and intra-individual variations are apparently taken into account by clinical trials which establish contraceptive effectiveness. The implication is that persons with the lowest bioavailability are adequately protected; by the same token those with average or increased bioavailability are in fact "overtreated," and there seems to be no way to fine-tune the oral contraceptive dose. The appearance of side-effects attributed to dosage may therefore be less a property of the formulation than of the bioavailability manifested by that particular individual, and there will inevitably be an irreducible minimum of certain adverse reactions.

Given this perspective of intra-inter-subject variance, one may examine the data of the multinational study carried out by Fotherby and his co-workers[6]. I emphasize this study, because of the consistency in dosage and experimental design, even though the number of subjects in each of the 14 centers was limited to 6 women. In our study[7] there was more variation in dose and the study of population ranged from 4 to 30 per center, with only 3 of the sets including women or more. In any event, the Fotherby study showed a wide range of uptake between and within centers and 24-hour plasma values which ranged from a low of 56 pg/ml in Alexandria to a high of 135 pg/ml in Bangkok, all after a standard 50 mcg. dose. The elimination rate constant also showed great variation; $t_{1/2}\beta$ ranged from 9 hr in London, Canada, to 31 hr in Sydney. Intersubject values also ranged widely: from 2.5 to 30 hr with a mean of 13.1 ± 7.8 hr. AUCs, as a measure of bioavailability, ranged from 5.8 to 12.5; however a low individual value of 2.25 ng·ml/hr was observed in Australia and a high of 29.4 in Brazil. Overall, as also noted in our own studies, there was a linear relation between C_{max} and AUC with R = 0.73.

With these profound intra- and inter-individual differences in pharmacokinetics, the reported differences in various populations would appear to be on somewhat uncertain ground, were it not for clear evidence of steroid-metabolic differences in these populations - differences which are consistent within these populations. In 1980 we reported[8] group differences in the nature of urinary conjugates of ethynyl estradiol as well as differences in urinary oxidative metabolite patterns. Examining chromatographic profiles of urinary conjugates, there were marked and consistent differences between Nigeria, Sri Lanka and the USA. Then, in examining the pattern of free steroids in the urine after hydrolysis, there were marked differences in the quantity of oxidative metabolites, chiefly the 2, 6 and 16-hydroxy compounds. Women in Nigeria showed the lowest excretion of oxidative metabolites; women in Sri Lanka showed substantially more, and women in the USA generally showed by far the most of these metabolites. The consistency of these findings ruled out intra-individual variability, thus supporting the concept of differences in the metabolism (and logically, therefore, in the pharmacokinetics) of ethynyl estradiol in different populations.

The proportion of circulating estrogen which is conjugated with glucuronic acid vs. sulfate must be the outcome of a very complex set of independent and interdependent reactions. While the molecular structure of the estrogen clearly has an effect, we know little of the other forces which play a role. They are certainly important in the present context, since the proportion of the two types of conjugates differed among populations. What nutritional factors might compete with the estrogens for the sulfatases is a question which derives directly from Dr. Back's fascinating studies of the competition between ascorbic acid and ethynyl estradiol. Except for our very preliminary examination of ethnic differences in the nature of the urinary conjugates of ethynyl estradiol, little work seems to have been done on this question.

The oxidative metabolism of the estrogens, which occurs generally at the 2, 6 and 16-positions, has been studied extensively and will not be reviewed. The important issue is the participation of the cytochrome P-450 group of enzymes, which are also involved in the metabolism of other drugs. As might be predicted, some of these drugs affect the plasma levels of ethynyl estradiol, while the reverse also occurs. Induction of these enzymes, for example by barbiturates, phenytoin and rifampicin might accelerate the metabolism and destruction of the estrogen. This has proved to be clinically significant only in the case of rifampicin. Other antibiotics such as the penicillins and tetracyclines are theoretical candidates for interaction, but the anecdotal data which are available are inconclusive. Enzyme-inhibitory compounds, such as MAO antidepressants, isoniazid and metronidazole might theoretically delay the degradation of the contraceptive steroid and increase the likelihood of side effects, but there is no clinical confirmation of this possibility.

It is reasonably well documented, however, that high-dose oral contraceptives diminish the clinical efficacy of certain analgesics such as acetaminophen, psychotropics such as imipramine and diazepam, bronchodilators such as aminophylline and theophylline.

All these interactions have been examined, albeit minimally in developed-world populations. Whether there are any clinically significant issues here for other populations with their individual characteristics is not known, and presents a formidable challenge in terms of the effort and cost involved. The fact that such issues have not surfaced in three decades of oral contraceptive use - most of it at what are currently called high doses - is superficially reassuring, but certainly no reason to ignore the problem.

References

1. Newburger J, Goldzieher JW. Pharmacokinetics of ethynyl estradiol: current view. *Contraception* 32:33. 1985.
2. Dyas J, *et. al.* A radioimmunoassay for ethinyl estradiol in plasma incorporating an immunosorbent, pre-assay purification procedure. *Ann Clin Biochem* 18:37. 1981.
3. Hümpel M, *et. al.* Investigation of pharmacokinetics of ethynylestradiol to specific consideration of a possible first-pass effect on women. *Contraception* 19:421. 1979.
4. Goldzieher JW, *et. al.* Human pharmacokinetics of ethynyl estradiol-3-sulfate and 17-sulfate. *Steroids* 51:63. 1988.
5. Brody SA, Turkes A, Goldzieher JW. Pharmacokinetics of three bioequivalent norethindrone/mestranol 50ug and three norethindrone/ethynyl estradiol 35ug formulations: are "low-dose" pills really lower? *Contraception* 40:269. 1989.
6. Fotherby K, *et. al.* Pharmacokinetics of ethynylestradiol in women from different populations. *Contraception* 23:487. 1981.
7. Goldzieher JW, Dozier TS, de la Pena A. Plasma levels and pharmacokinetics of ethynyl estrogens in various populations. I. Ethynyl estradiol. *Contraception* 21:1. 1980.
8. Williams MC, Goldzieher JW. Chromatographic patterns of urinary ethynyl estrogen metabolites in various populations. *Steroids* 36:255. 1980.

It appears that both pharmacokinetic and steroid-metabolic studies are required for a definitive insight into the characteristics of inter-population differences.

PHARMACODYNAMICS OF NORPLANT[R] IMPLANTS

Harold A. Nash and Dale N. Robertson[1]

1. INTRODUCTION

NORPLANT[R] subdermal implants comprise a contraceptive modality in which levonorgestrel is continuously released from capsules placed beneath the skin. The capsules are made of a dimethylsiloxane/methylvinylsiloxane copolymer (Silastic[R]) and are filled with crystalline levonorgestrel. Each capsule is of 34 mm total length with 30 mm of the length being filled with steroid. The diameter is 2.4 mm. Six capsules are placed in a fan-shaped pattern through a single 3 mm incision. Clinical studies have shown effectiveness through five years.

2. METHODS

The dose of levonorgestrel supplied by the implants has been estimated by analyzing implants recovered from women for steroid remaining after various periods of use. The slopes of regression curves for steroid content as a function of time of use allow estimation of the daily dose. Further information on the pattern of release during use has been obtained from changes in levonorgestrel concentrations in plasma or serum during use. The measurements of plasma concentrations were all carried out in the same laboratory using the radioimmunoassay system described by Weiner and Johansson[1]. Measurements of serum concentrations were similarly carried out in a single laboratory using the radioimmuno-

H. NASH and D. ROBERTSON • The Population Council, Center for Biomedical Research, 1230 York Avenue, New York, NY 10021

assay methods of Stanczyk *et al*[2]. Except as specifically noted, circulating concentrations quoted in this paper have been obtained by the methods of Weiner and Johansson[1].

Information on potential release rates has been obtained by measuring *in vitro* release from new and recovered implants. These measurements were made using 1:750 aqueous benzalkonium chloride solution as the bathing medium. The benzalkonium chloride enhances wetting of the capsule surface and prevents microbial growth. One end of a capsule was fixed to the bottom of a 20 ml vial with Silastic Medical Adhesive and 15 ml of the benzalkonium chloride solution introduced. The vial was placed on its side in a 37-degree C. water bath with shaking at 100 one-inch strokes per minute. The bathing solution was changed daily and its levonorgestrel content assayed by HPLC.

3. RESULTS

3.1 Release rate in vivo

The amount of steroid found in new and recovered implants is represented graphically in **Figure 1A**. To aid in estimating steroid release rates, over-lapping, time-limited regression curves have been calculated. The curves were constructed for 2-year periods beginning at successive six-month intervals, except a separate curve was calculated for the first 6 months and the last three curves end at 61 months, with the last two beginning at 42 and 48 months. The family of curves is shown graphically in **Figure 1B**. The family of curves was used to draw a smoothed curve and tangents at selected points used to calculate mean release rates. The rates so calculated are shown in **Table 1**.

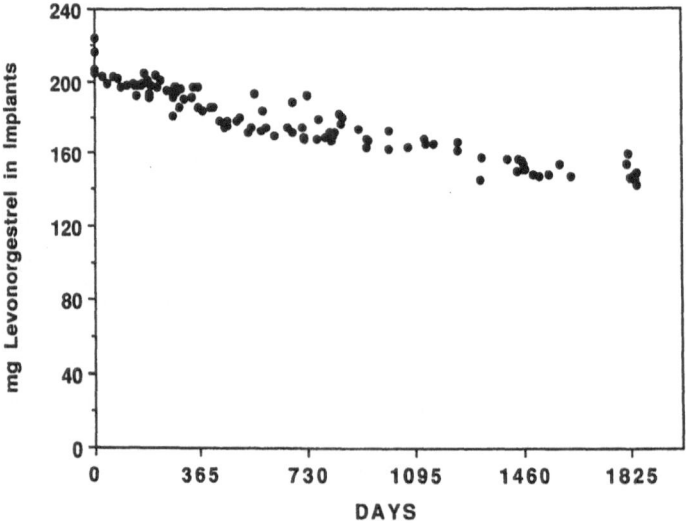

Figure 1A. Levonorgestrel remaining in sets of implants after periods of use.

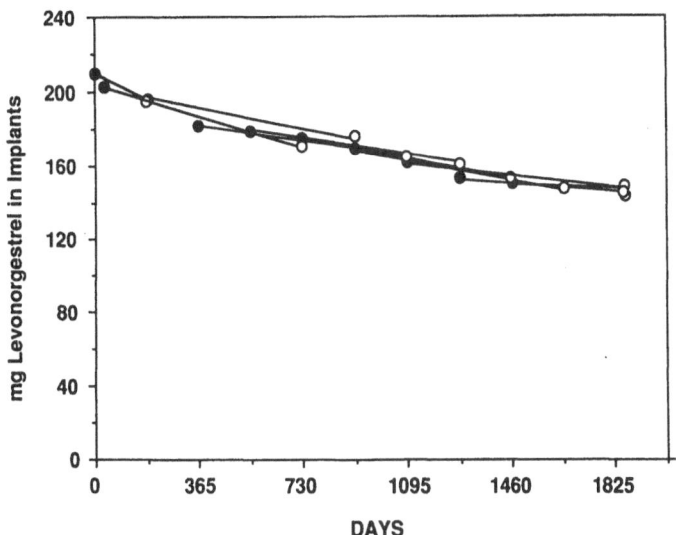

Figure 1B. Regression Lines: Levonorgestrel remaining in sets of implants after periods of use.

Table 1. Release rates from Norplant^R implants as calculated from analysis of recovered implants.

Months	µg/day
3	75
9	53
18	36
30	30
>30	29

Table 2.   Plasma levonorgestrel
concentrations calculated from regression curves and normalized to
60 kg body weight.

Month	Levonorgestrel concentrations ng/ml
12	.326
24	.306
36	.286
48	.266

3.2 Blood Levels

Changes in blood levels over time may be assumed to reflect dosage, except in the weeks before concentrations of sex hormone binding globulin have stabilized. Mean blood level concentrations among NORPLANT[R] users as calculated from regression curves and normalized to 60 kg body weight are shown in **Table 2**.

Concentrations have proven to be importantly affected by body weight as illustrated by the curves of mean concentrations for groups of differing body weight shown in **Figure 2**[3]. The mean change with change in body weight as determined in multiple regression analysis of levels between months 4 and 63 is a decrease of about 3.3 pg/ml for each kg increase in body weight. The rapid initial decrease in circulating concentrations evident in **Figure 2** is at least partly a reflection of the role of sex hormone binding globulin (SHBG) in carrying levonorgestrel in circulation and the role of levonorgestrel in depressing SHBG concentrations. The decrease in SHBG during early days of NORPLANT[R] use is reflected in **Figure 3**, reproduced from a publication by Affandi *et al*[.]. The same authors have reported significant correlations between SHBG levels and levonorgestrel levels for each interval examined during 60 months of NORPLANT[R] use.

Although mean circulating concentrations provide overall patterns, they provide little basis for predicting circulating concentrations in an individual user. This is illustrated by regression curves for 21 NORPLANT[R] users in Uppsala selected for regression analysis because they had at least five levonorgestrel assays with at least one beyond 36 months of use. The individual curves are reproduced graphically in **Figure 4**. All except four are negative in slope as would be predicted from mean concentrations among groups of

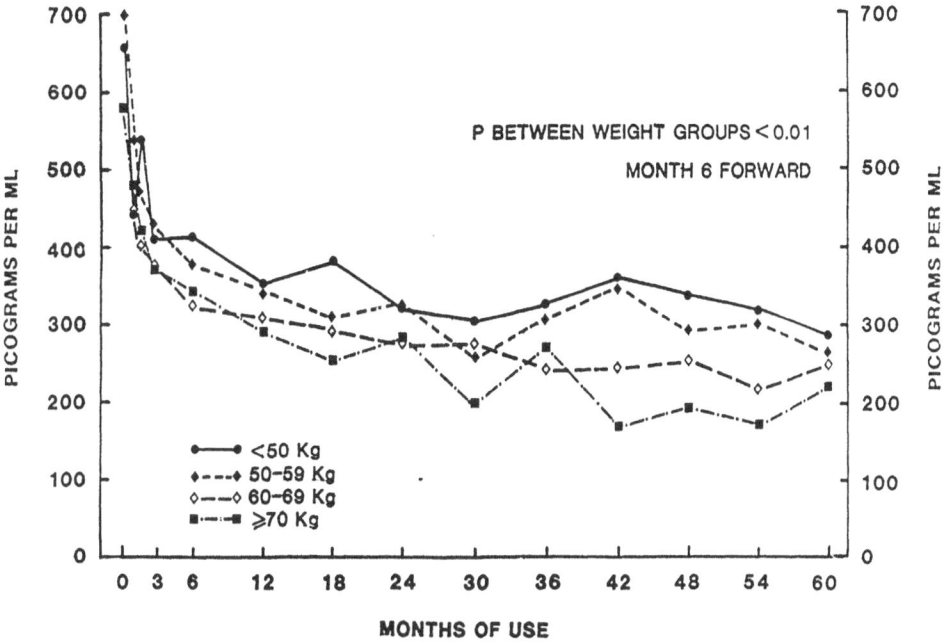

Figure 2. Norplant[R] Capsules: Plasma concentrations of levonorgestrel(pg/ml) by weight at admission and by time at months of sampling. Uppsala assay.

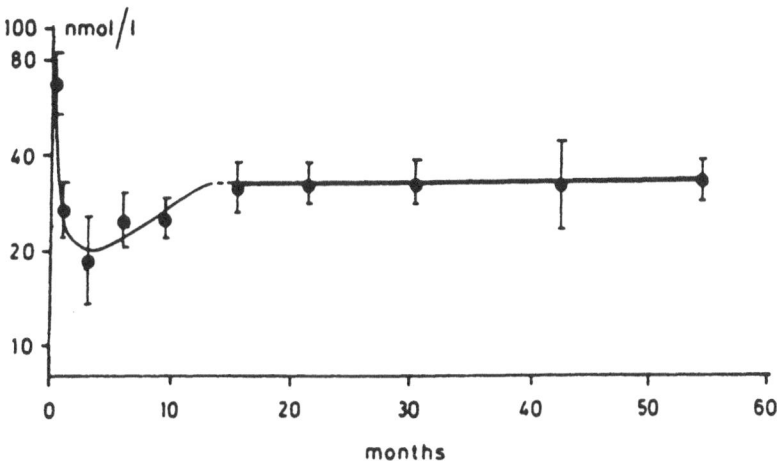

Figure 3. Geometric means and 95% confidence limits for SHBG levels in Norplant[R] users[4].

women as a function of time of implant use. The differences in circulating levonorgestrel concentrations among these 21 women are as great as 3.5 fold. It is also to be noted from the regression curves that women tend to retain the same rank order over time. Those who have initially high concentrations tend to remain high and those with initially low concentrations tend to remain low. The differences in levonorgestrel concentrations among these 21 women is only partially accounted for by differences in body weight evident in the association between concentrations and body weight plotted in **Figure 5**.

The levonorgestrel concentrations measure immediately after implant placement can give some information on the maximum dose released following insertion. The smoothed curve for dose calculated from analysis of recovered implants indicated a dose of about 31 ug/day at 24 months. The mean blood level at that time in Santiago clinic was 0.28 ng/ml. The blood level at one week in studies conducted in Santiago, Chile was 0.74 o.17 (N=9) ng/ml or 2.6 times that at 24 months. Assuming a straight-line relationship between dose and blood level, one can multiply 31 ug by 2.6 to arrive at an estimated dose of 81 ug/day during the first week. Considering the higher SHBG levels at one week than at 24 months, the actual dose was probably lower.

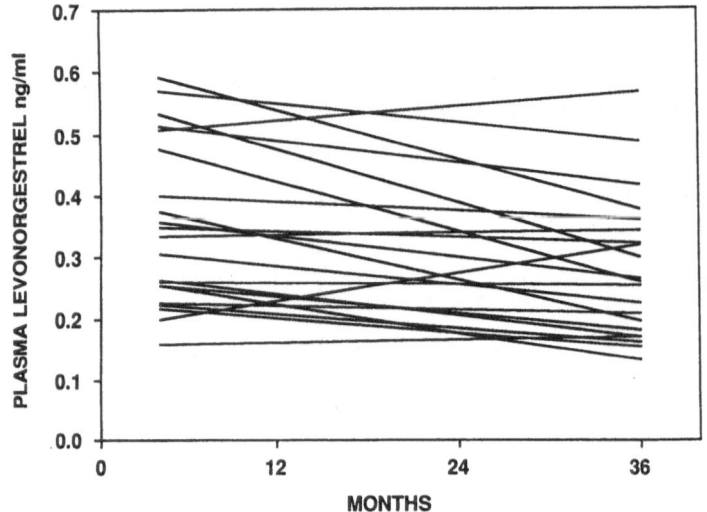

Figure 4. Plasma levonorgestrel regression for individual Norplant[R] users. Uppsala, Sweden.

Examination of the blood levels required for effectiveness allows for some broad statistical generalizations but provides little basis for predicting the risk of pregnancy in individual women. The broad generalizations are illustrated by the relationships between blood levels and weight (**Figure 2**) and pregnancy rates and weight. The relationship between body weight and pregnancy rates are indicated in **Table 3**. The blood level data and the pregnancy data are not directly interrelated in that the blood level data is confined to implants made with "soft" tubing while the pregnancy data includes both "soft" and "hard"

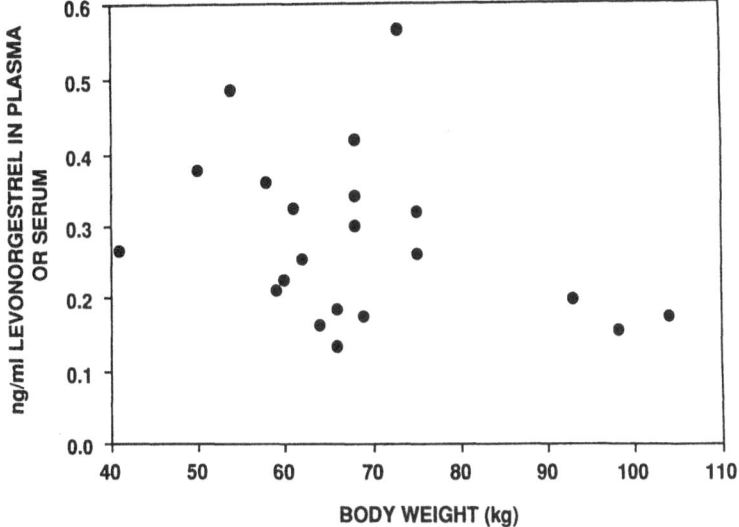

Figure 5. Levonorgestrel concentration at 36 months as a function of body weight.

Table 3. Cumulative gross pregnancy rates at the end of 5 years as a function of body weight.

Weight Class-kg	N	Pregnancy rate
<50	532	0.2
50-59	1046	3.4
60-69	587	5.0
≥70	303	8.5

tubing[a]. There were not enough pregnancies with the "soft" tubing to illustrate the point.

To examine the predictability for individual women, probable circulating concentrations at the time of conception were estimated for women who became pregnant on levonorgestrel implants and who had levonorgestrel concentration measurements at times reasonably close to the time of conception. These concentrations were compared with the mean concentrations among users in the same clinic after equal periods of use. The comparisons are shown in **Table 4**.

---

[a]The difference between the "soft" and "hard" tubing lies in the amount of silica filler. Differences in feel and elasticity are small.

The mean concentration associated with pregnancy was not markedly below the mean at equal times of use for the clinic in which it occurred. Among 25 pregnancies for which comparisons can be made between estimated concentrations at conception and the pertinent clinic means, six (24 percent) had estimated concentrations at conception slightly to moderately above the clinic mean. Among 135 women from the clinics using implants of one production lot and whose plasma samples were assayed in the Uppsala laboratory, 52 percent had one or more values below 0.25 ng/ml and 21 percent had values below 0.20 ng/ml. In spite of this frequency of values in a range associated with pregnancies, there were only two pregnancies among women using this lot in those clinics.

The effects of some other factor on levonorgestrel pharmacokinetics have been explored. Implants have been placed both in the palmer aspect of the forearm and in the inner side of the upper arm. Neither mean circulating concentrations, nor mean steroid loss after equivalent periods of time indicate any difference in steroid dose between these

Table 4. Mean estimated plasma or serum levonorgestrel concentrations at conception and after equal periods of use. Norplant[R] and Norplant[R]II.

Levonorgestrel Assay site[a]	Mean levonorgestrel concentrations-ng/ml+-SD at conception	for clinic
Uppsala	0.21±0.09[b]	0.26±0.06
Los Angeles	0.28±0.10[c]	0.34±0.02

a. Refers to site of levonorgestral assay –
   not necessarily to the site of the pregnancy.

b. N=15   c. N=11

placement sites. Exogenous estrogens increase blood levels through their effect in increasing SHBG[5]. Two drugs which increase liver microsomal activity, phenytoin and carbamazepine, decreased plasma levonorgestrel concentrations.

3.3 In vitro release rates

Measurement of *in vitro* release rates has been conducted on new implants and on implants recovered from women. There are progressive decreases in *in vitro* release rates as a function of time of use, with the more rapid decreases occurring within the first year of use. Mean release rates for sets of new and recovered implants are shown in **Figure 6**. The apparent relationships between *in vitro* release and release during residence *in vivo* is shown in **Figure 7**.

4. DISCUSSION

It is relatively difficult to attain the precision one would like in assessing the pharmacokinetics of a long lasting method such as NORPLANT[R]. Analysis of recovered implants can give a fairly accurate estimate of the mean dose delivered over the period of use but does not give precise information on the dose profile over time. Blood levels over time allow for some refinement of the pattern. For example, the data in **Figure 1** hint at a substantial slowing of release between 48 and 60 months. Blood level profiles (**Figure 2**) indicate this is unlikely. If one can assume that clearance rates will remain constant, more refined estimates of release rate patterns could be attained by measuring levonorgestrel clearance rates before implant placement, thus providing a basis from blood levels for calculating dose being received. Uncertainty would, however, be introduced by changing SHBG concentrations.

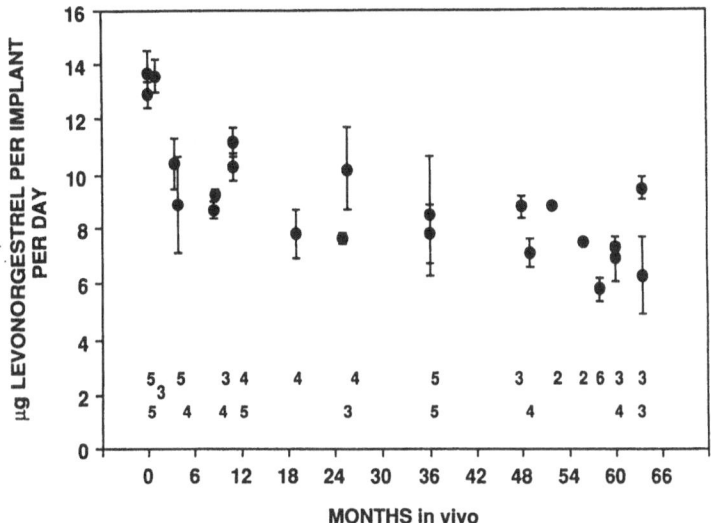

Figure 6. *In vitro* levonorgestrel release rates of sets of implants recovered from subjects.

The scatter of points in **Figure 1A** can be interpreted as differences in release rates among women. The magnitude of the scatter is indicated by the fact that the standard deviation about the smoothed curve in the third year of implant use is the equivalent of ±5.8 ug/day during the course of implant use. The standard deviation in the original load of steroid in the implants was large enough to account for this degree of scatter. However, experience with other manufacturing lots, not included in the data in this report because the Silastic[R] was of slightly different composition, indicates that part of the variation does appear to represent real differences in dose provided. The steroid load in those implants was individually weighed during filling to a tolerance of ±2 mg (equivalent to ±1.6 ug/ day over 3.5 years). The standard deviation between sets in the third year was equivalent to ±5.6 and ±4.2 ug/day. The expected variation among sets as predicted from the variation among implants in individual sets was of the order of ±1.0 ug/day over 3.5 years. The cause of the variation in release among women may lie in varying amounts of fibrous capsule formed around the implants.

Several factors may contribute to the variation in blood levels. Differences in release rate among implant sets are one factor. Differences in metabolic clearance rates are undoubtedly another factor. Hümpel et al[6] have reported differences as great as 5 fold in clearance among women receiving an acute dose of levonorgestrel *i.v.* and as great as 4 fold among women receiving a combination pill containing levonorgestrel. Other investigators reported much less variation. The coefficient of variation in metabolic clearance rates of levonorgestrel and estradiol orally in studies by El-Raghy et al[7] was 0.19 to 0.24. Similar variation was reported by Back et al[8].

The decrease in levonorgestrel release in the first year partly reflects the decrease in release capability as evidenced in decreasing *in vitro* release rates but the *in vivo* release soon declines to levels well below the *in vitro* release. While it is true that the *in vitro* system is an artificial system that would not be expected to accurately predict *in vivo* release, changing relationships between *in vivo* and *in vitro* release must signal the intervention of new phenomena. One possibility is the modulation of release by the thin fibrous capsule that forms around the implants. Differing fibrous capsule thickness may also account for some of the difference in release rate among individuals, although this has not been proven. The decrease in *in vitro* release has been shown to be partially a function of decreased efficiency of wall contact by the crystals.

Attempts to correlate blood levels among NORPLANT[R] users with performance other than pregnancy rates and ovulation incidence have been only moderately fruitful. Inspection of side effects over time shows the number of onsets of bleeding or spotting as analyzed from menstrual diaries to parallel the declining blood level curve, but the chance that this is coincidence seems high. Some correlation has been found between weight and bleeding pattern and therefore between blood levels and bleeding pattern. Higher weight women have experienced more bleeding days[3].

Some of the answer to effects of varying doses derives from two studies with covered rods delivering levonorgestrel doses approximately 1.5 and 2.25 times those delivered by NORPLANT[R]. Mean blood levels at 12 months were 1.1 and 2.4 times as high as among NORPLANT[R] users in the same clinic. At 36 months the ratios were 1.4 and 1.9. The studies were of moderate size - 78 and 111 acceptors - so don't allow of fine distinctions. Ovulatory cycles as judged by progesterone peaks were markedly reduced. Bleeding patterns were shifted toward markedly less bleeding with approximately 55 and 90 percent of subjects experiencing 60 or more days without bleeding in three months reference periods in the first year as compared with about 35 percent among NORPLANT[R] users. In collecting information on other side effects, all complaints recorded by the physician in response to the general question, "How have you been feeling since the last visit?" asked at each follow-up visit and all conditions diagnosed by the clinic were recorded and analyzed. This resulted in entries in 169 categories in the NORPLANT[R], NORPLANT[R]II or high dose studies. There were few indications of dose-related differences in side effect incidence. Nervousness and anxiety were more frequent complaints among users of the higher doses but not in a dose-related manner. Headache

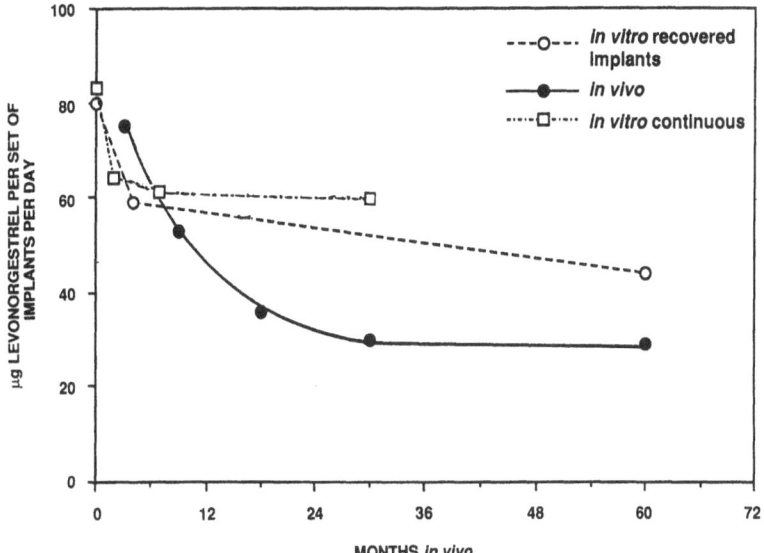

Figure 7. *In vitro* and *in vivo* release from sets of implants.

was also somewhat more frequently reported in the studies utilizing the higher doses and particularly at the highest dose. Weight gain was a more frequent complaint at the highest dose than in other studies, but the lowest frequency of complaint of weight gain was at the intermediate dose. Complaints of breast discharge were particularly frequent at the intermediate dose, but were relatively infrequent at the highest dose. The same relationships were true of leukorrhea and mastalgia. Complaints of amenorrhea were very frequent at the highest dose and complaints of excessive bleeding were low relative to the other dose levels.

References

1.   Weiner, E. and Johansson, E.D.B. Plasma levels of d-norgestrel, estradiol, and progesterone
      during treatment with Silastic implants containing dnorgestrel. Contraception 14:81-92 (1976).
2.   Stanczyk, F.Z., *et al*. Radioimmunoassay of serum d-norgestrel in women following oral and
      intravaginal administration. Contraception 12:279-298 (1975).
3.   Sivin, I. Unpublished.
4.   Affandi, B., *et al*. The interaction between sex hormone binding globulin and levonorgestrel
      released from NORPLANT[R], an implantable contraceptive. Contraception 35:135-144 (1987).
5.   Weiner, E. and Johansson, E.D.B. Contraception with d-norgestrel Silastic rods. Plasma levels
      of d-norgestrel and influence on the ovarian function. Contraception 14:551-562 (1976).
6.   Hümpel, M., *et al*. Investigations of the pharmacokinetics of levonorgestrel to specific
      consideration of a first pass effect. Contraception 17:207-220 (1978).
7.   El-Raghy, I., *et al*. Pharmacokinetics of oral contraceptive steroids in Egyptian women: studies with
      oral NORDETTE and NORMINST. Contraception 33:379-384 (1986).
8.   Back, D.J., *et al*. The pharmacokinetics of levonorgestrel and ethynylestradiol in women -
      Studies with Ovran and Ovranette. Contraception 23:229-239 (1983).

PHARMACOKINETIC DRUG INTERACTIONS WITH ORAL CONTRACEPTIVES

D.J. Back and M.L'E. Orme[1]

SUMMARY

Over the last 20 years there have been a number of case reports and clinical studies on the topic of drug interactions with oral contraceptive steroids (OCS). Some of the documented interactions are of definite therapeutic relevance whereas others can be discounted as being of no clinical significance. Pharmacological interactions between OCS and other compounds may be of two kinds. 1. Drugs may impair the efficacy of OCS leading to breakthrough bleeding and pregnancy. In a few cases OCS activity is enhanced. 2. OCS may interfere with the metabolism of other drugs.

A number of anticonvulsants (phenobarbitone, phenytoin, carbamazepine) are proven enzyme-inducing agents and thereby increase the clearance of the OCS. In contrast, sodium valproate has no enzyme-inducing properties and thus women on this anticonvulsant can reasonably rely on their low dose OCS for contraceptive protection. In the past couple of years the molecular basis of this interaction has emerged, with evidence of specific forms of cytochrome P450 (P45011C and 111A gene families) being induced by phenobarbitone. Rifampicin, the antituberculous drug, also induces a cytochrome P450 which is a product of the P450lllA gene subfamily. This isozyme is one of the major forms involved in ethinylestradiol (EE_2) 2-hydroxylation.

D. BACK and M. ORME • The University of Liverpool, Department of Pharmacology and Therapeutics, New Medical Building, Ashton Street, Liverpool L69 3BX, United Kingdom

Broad spectrum antibiotics have been implicated in causing pill failure; case reports document the interaction and family planning doctors are convinced that the interaction is real. The problem remains that there is still no firm clinical pharmacokinetic evidence which indicates that blood levels of OCS are altered by antibiotics. However, perhaps this should not surprise us given that the incidence of the interaction may be very low. It is our belief that a woman at risk will have a low bioavailability of EE_2, a large enterohepatic recirculation and a particularly susceptible gut flora to the antibiotic being used.

Although early studies suggested that vitamin C (ascorbic acid) gave rise to increased blood concentrations of EE_2 due to competition for sulphation, more recent work has cast doubt on this. On theoretical grounds adsorbents (e.g. magnesium trisilicate, aluminum hydroxide, activated charcoal and kaolin) could be expected to interfere with OC efficacy, but there is no firm evidence that this is the case. Similarly, there is no evidence that smoking alters OCS pharmacokinetics.

OCS are now well documented as being able to alter the kinetics of other concomitantly administered drugs. The clearance of a number of benzodiazepines undergoing oxidation (chlordiazepoxide, diazepam) and nitro reduction (nitrazepam), theophylline, prednisolone, caffeine and cyclosporin is reduced in OCS users. The clearance of some drugs undergoing glucuronidation (temazepam, salicylic acid, paracetamol, morphine, clofibric acid) is apparently increased.

Finally, despite a recent report that the progestogen component present in an OC formulation can alter the disposition of EE_2 other studies have failed to confirm this finding.

INTRODUCTION

Oral contraceptive steroids continue to be the most popular form of reversible contraception in most countries of the world. One thing is certain: that the majority of the women on OCS will at some stage be taking other drugs. There are two general categories of drug interactions involving OCS to consider: 1) those where the interacting drug alters the disposition of the OC. Most important is where the blood levels of the OC are reduced thereby leading to impaired efficacy, breakthrough bleeding and occasionally pregnancy. In a few cases OCS activity is enhanced. 2) OCS may also interfere with the metabolism of other compounds.

The subject of OC-drug interactions has been reviewed many times (Back et al. 1981a, 1983; Back & Orme, 1990; Fraser & Jansen, 1983; Orme, 1982; Orme & Back, 1980; Roberton & Johnson, 1976; Shenfield, 1986; Stockley, 1979; Szoka & Edgren, 1988). The aim of the present review is to give a critical analysis, and where it is known a mechanistic basis of the interactions.

1. DRUGS INTERFERING WITH ORAL CONTRACEPTIVE EFFICACY

1.1 Anticonvulsant Drugs

A wide variety of anticonvulsant drugs have been implicated as causing breakthrough bleeding or failure of contraception in women taking OCS. This was first reported by Kenyon (1972) but other cases have been reported (Belaisch et al., 1976; Coulam & Annegers, 1979; Diamond et al., 1985; Gagnaire et al., 1975; Janz & Schmidt, 1974; Hempel et al., 1973). The drugs implicated include phenytoin, phenobarbitone, methylphenobarbitone, primidone, carbamazepine and ethosuximide. Phenytoin appears to be the most commonly implicated anticonvulsant and this is born out by data from the United Kingdom Committee on Safety of Medicines (CSM).

Between 1973 and 1984 a total of 43 cases of contraceptive failure in women taking anticonvulsants with their OCS therapy were reported to the CSM (Back et al, 1988, Table 1). Phenytoin accounted for 25 of these cases, phenobarbitone for 20 with relatively smaller numbers for other anticonvulsants. The CSM monitors adverse drug reactions by means of the yellow card reporting system. However, Lumley et al. (1986) indicated that less than 10% of adverse reactions are actually reported to the CSM; the main reasons being the reaction was too trivial, was already well known, or there was uncertainty of causal relationship. If the same degree of under-reporting applies to the OC-drug interactions then the 43 pregnancies in women concurrently taking anticonvulsant drugs is a very inadequate estimate of the number which have actually occurred. It should be noted, however, that the reported data tell us nothing of the actual prevalence of the alleged interactions. We do not have good data for unintended pregnancy in women who were (a) not receiving any other medication or (b) receiving a drug not implicated in an interaction.

The clinical pharmacokinetic data of this interaction is relatively sparse. In one study (Back et al., 1980a) phenobarbitone was given in doses of 30 mg b.d. to 4 young women with epilepsy. They were followed over a complete cycle of OCS use before phenobarbitone was commenced and then for 2 cycles after phenobarbitone was started. Two out of the 4 women showed a significant fall in steady state ethinylestradiol (EE_2) concentrations and both had breakthrough bleeding; however, the other 2 women showed no change in blood level. Although there was no significant change in the plasma concentration of the progestogen norethisterone, there was a significant increase in sex hormone binding globulin (SHBG) capacity. Such an increase effectively reduces the free progestogen concentrations.

Pharmacokinetic data from other studies (Crawford, 1986, 1989) involving single dose administration of an OCS (50 ug EE_2, 250 ug levonorgestrel) to women before and 8 to 12 weeks after starting anticonvulsant therapy for tonic clonic seizures are shown in Figure 1 and Table 2. In the case of phenytoin there was a significant decrease in the area under the plasma concentration-time curve (AUC) for both EE_2 and levonorgestrel (LNG). Similarly in all patients receiving carbamazepine there was a marked decrease in AUC of both EE_2 and LNG. The degree of change is quite likely to be sufficient to produce contraceptive failure in women on long term OCS. Sodium valproate in contrast to phenobarbitone, phenytoin and carbamazepine had no detectable effect on the kinetics of EE_2

Table 1. Anticonvulsant drugs involved in alleged inter-actions

Anticonvulsant drugs	Number of Reports
Phenytoin	25
Phenobarbitone	20
Primidone	7
Carbamazepine	6
Ethosuximide	4
Sodium valproate	1

The total is greater than 43 because of multiple drug therapy.

and LNG (Table 2). While the latter drugs are known to be enzyme inducing agents in man, sodium valproate has no enzyme inducing action. Based on these data it is possible to state that women taking sodium valproate can rely on conventional low dose OCS for contraceptive control, whereas women receiving phenobarbitone, phenytoin or carbamazepine should initially use a higher dose OCS if they wish to rely for contraception on their OCS preparation.

In order to understand the molecular basis of the interaction it is necessary to examine the mechanism of increased metabolism of OCS, particularly EE_2. A major pathway of metabolism of EE_2 is 2-hydroxylation by enzymes of the cytochrome P450 superfamily. Guengerich (1988) has argued that the specific isozyme responsible for EE_2 2-hydroxylation is P450lllA4. Evidence for this conclusion included results of studies using enzyme reconstitution, immunoinhibition, correlation of activity and inhibitors. The working hypothesis then is that phenobarbitone and other anticonvulsant drugs are able to induce this isozyme and hence increase the rate and extent of 2-hydroxylation. Additional studies in the same area have been performed by Ball et al. (1990). Sixteen human livers were examined for a variety of enzyme activities including EE_2 2-hydroxylase. The highest EE_2 2-hydroxylase activity was found in the liver of a subject who had received phenobarbitone and

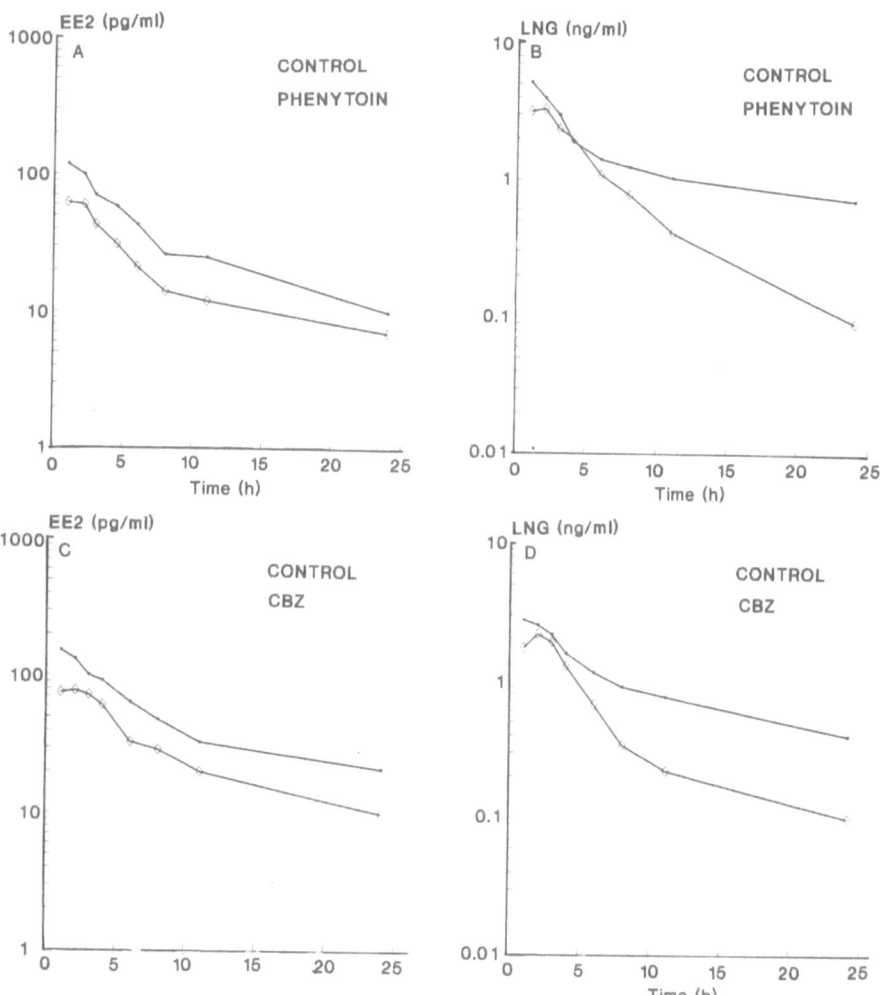

Figure 1. Plasma concentration-time profiles of ethinylestradiol (EEs) and levonorgestrel (LNG) in patients taking either phenytoin (a&b) or carbamazepine (CBZ, c&d).

Table 2. Area under the plasma concentration time curve (AUC_{0-24}) for ethinylestradiol and levonorgestrel in patients taking either phenytoin, carbamazepine or sodium valproate.

Patient No.	Daily dose of Antiepileptic (mg)	Ethinyloestradiol AUC (pg.ml⁻¹.h)		Levonorgestrel AUC (ng.ml⁻¹.h)	
		Control	Test	Control	Test
Phenytoin					
1	250	641	303	41.8	22.6
2	200	956	326	23.1	17.2
3	200	944	234	10.5	13.7
4	200	785	1060	18.9	5.9
5	300	770	332	46.7	25.4
6	300	740	208	60.8	32.2
Mean ± S.D.		806 ± 122	411* ± 132	33.6 ± 19.2	19.5* ± 9.3
Carbamazepine					
1	400	1833	630	14.9	10.5
2	300	1024	960	20.2	8.7
3	600	750	451	20.1	14.3
4	600	1047	648	36.6	21.9
Mean ± S.D.		1163 ± 466	672* ± 211	22.9 ± 9.4	13.8* ± 5.8
Sodium Valproate					
1	400	788	964	35.2	28.1
2	400	969	1075	20.9	31.5
3	400	981	873	26.4	33.1
4	400	758	1112	25.9	25.3
5	400	1339	1399	-	-
6	400	542	441	-	-
Mean ± S.D.		880 ± 267	977 ± 319	29.1 ± 5.8	29.2 ± 3.8

*, $P < 0.05$

phenytoin for more than 25 years. In other studies antibodies raised in rabbits to rat hepatic cytochrome P450 isozymes were used to inhibit EE_2 2-hydroxylase activity of human liver samples. Possible correlations between the proteins recognized by the antibodies and 2-hydroxylation rate were determined. These experiments provide evidence that the 2 hydroxylation of EE_2 is catalyzed by cytochromes from the P450IIIA gene subfamily (in agreement with Guengerich, 1988) and in addition by isozymes from the P450IIC and IIE subfamilies; some of these proteins are inducible by the anticonvulsants mentioned above.

1.2 Antibiotics (excluding rifampicin)

Broad spectrum antibiotics have been implicated in causing failure of contraception in women on the pill. There are a number of case reports (Bacon & Shenfield, 1980; Bainton, 1986; De Sano & Hurley, 1982; Dossetor, 1975) and Back et al. (1988) listed a total of 63 pregnancies reported to the CSM (Table 3). In this study, the most widely prescribed antibiotics (penicillins and tetracyclines, comprising 74% of all antibiotic prescriptions in England) were implicated in the majority (70%) of pregnancy reports. In a study of pill method failures in reliable pill takers, documented over a 4-year period (1981-85) in New Zealand by Sparrow (1987), 23% of the 163 cases were stated as being associated with antibiotics. In addition, family planning doctors have no doubts that such interactions occur. It is, therefore, important to ascertain why clinical pharmacokinetic studies have been singularly unsuccessful in demonstrating any consistent effect of antibiotics on plasma concentrations of synthetic steroids (Friedman et al., 1980; Joshi et al., 1980; Back et al., 1982c).

From early studies in pregnant women the view emerged that interference in steroid metabolism was related to antibiotics acting on gut microflora (Boehm et al., 1974). There is good evidence that EE_2 undergoes enterohepatic recirculation (EHC). EE_2 can be conjugated with both sulphate and glucuronic acid. Sulphation occurs primarily in the small intestinal mucosa (Back et al., 1981b; Back et al., 1982b; Rogers et al., 1987a) while conjugation with glucuronic acid occurs mainly in the liver (Helton et al., 1976; Salhberg et al., 1981). These conjugates are excreted in the bile and thus reach the colon where they may be hydrolysed by the bacteria present (principally Clostridia spp, Chapman, 1981) to liberate unchanged EE_2 which can be reabsorbed into the portal circulation. It is difficult to find out how important the EHC of EE_2 is in practical terms. There have been reports in the literature of secondary peaks in plasma concentration time profiles which are presumably due to EHC (Back et al., 1979a).

Oral antibiotics kill off the clostridia (and other bacteria) responsible for the hydrolytic process. Thus any conjugates of EE_2 present in the lower part of the gastrointestinal tract are lost in the faeces since they are too hydrophilic to be absorbed themselves. In animal studies the EHC of EE_2 is extensive and is clearly diminished by treatment with a variety of antibiotics (Back et al., 1982a). The degree of reduction of EHC can be correlated with the degree of suppression of the gut microflora. However it has been difficult to duplicate such studies in humans. Thus ampicillin (Back et al., 1982c; Friedman et al., 1980; Joshi et al., 1980) had no effect on the kinetics of EE_2 in women even though the ability of faecal micro organisms to hydrolyse steroid conjugates is dramatically reduced (Chapman 1981). Similarly, cotrimoxazole and tetracycline did not cause any reduction of plasma steroid concentrations (Orme & Back, 1986; Grimmer et al., 1983). More recently we have failed to show any alteration of OC disposition when women were taking the quinolone antibiotic temafloxacin or the macrolide antibiotic clarithromycin (Orme et al., 1990a,b). These studies are summarized in Table 4.

Table 3. Antibiotics involved in alleged interactions reported to the CSM 1968-84.

Antibiotic	Number of Reports
Penicillins	32
Tetracyclines	12
Cotrimoxazole	5
Metrinidazole	3
Cephalosporins	2
Trimethoprim	2
Erythromycin	2
'Antibiotic'	2
Sulphonamides	1
Griseofulvin	1
Fucidic Acid	1

Although these studies have failed to show any systemic interaction between antibiotics and OCS it is very possible that there are certain individuals who may be at risk of this interaction. Such individuals are most probably (but not exclusively) women with a low bioavailability of EE_2 due to extensive steroid metabolism in the gut wall and liver, a large EHC of EE_2 and a particularly susceptible gut microflora to the antibiotic being used. It is worth speculating that if a woman was unable to hydroxylate EE_2 (i.e. was deficient in the relevant cytochrome P450 isozyme(s)), she may form EE_2 conjugates to a greater extent and hence have a particularly large EHC. Maggs et al. (1983) identified one woman who was unable to 2-hydroxylate EE_2 (as judged by biliary metabolites).

It is impossible at the present time to evaluate fully the influence of antibiotics on steroid kinetics and different physicians may well offer different advice. One objection to some of the reported studies is that the bacteria in the gut develop resistance to antibiotics fairly rapidly. If the antibiotic interaction is looked for after 2 weeks therapy with the antibiotic it may be missed, since resistance to ampicillin develops over 5 to 6 days (Leigh

Table 4. Clinical pharmacokinetic studies involving antibiotics and OCS which have failed to show any inter-action.

Antibiotic	Number of Women	Reference
Ampicillin (500 mg t.i.d. for 7 days)	13	Back et al 1982c
Ampicillin (250 mg q.i.d. for 16 days)	11	Friedman et al 1980
Ampicillin (500 mg b.i.d. for 5-7 days)	6	Joshi et al 1980
Cotrimoxazole (320/1600* mg daily for 7 days)	9	Grimmer et al 1983
Tetracycline (250 mg b.d. for 4 weeks)	5	Orme & Back 1986
Temafloxacin (600 mg b.d. for 7 days)	12	Orme et al 1990a
Clarithromycin (250 mg b.d. for 7 days)	10	Orme et al 1990b

*Total daily dose of 320 mg trimethoprim
             1600 mg sulphamethoxazole

et al., 1976). However, several of the studies mentioned above (Back et al., 1982c; Grimmer et al., 1983) looked at the interaction during a 7-day course of the antibiotic starting 3 days after the commencement of the antibiotic therapy at a time when resistance would be unlikely to have developed, and still no interaction was detected.

1.3 Rifampicin

The original report of an interaction between rifampicin and OCS was by Reimers & Jezek (1971). Numerous other reports appeared during the 1970s and these were reviewed by Orme et al. (1983). Rifampicin is a potent enzyme inducing agent and in a clinical pharmacokinetic study in 8 patients, ethinylestradiol and norethisterone plasma concentrations were determined both during a course of rifampicin therapy and again 1 month after stopping the rifampicin (Back et al., 1979b, 1980c). Rifampicin therapy caused a reduction in norethisterone AUC from 37.8 to 21.9 $ng.ml^{-1}.h$ ($P < 0.01$) and of EE_2 AUC from 1749 to 1015 $pg.ml^{-1}.h$ ($P < 0.05$). All patients showed evidence of microsomal enzyme induction as judged by an increase in the urinary excretion of 6B-hydroxycortisol and an increased plasma clearance of antipyrine.

There is recent evidence (Guengerich, 1988; Combalbert, 1989) that rifampicin induces cytochrome P450lllA4, one of the major enzymes involved in EE_2 2-hydroxylation. In the study of Combalbert et al. (1989), enzyme activity towards the calcium antagonist nifedipine was shown to be clearly correlated with the level of P450lllA4 determined from Western blots, and the enzyme activity and hepatic levels of protein were simultaneously increased in the livers of patients treated for 4 days with 600 mg rifampicin per day. Since Guengerich (1988) has indicated that nifedipine and EE_2 are metabolized by the same P450 isozyme we can, by inference, see a molecular basis for the rifampicin - OCS interaction.

1.4 Griseofulvin

A report by van Dijke & Weber (1984) recorded 20 cases of women on long term OCS who experienced intermenstrual bleeding or amenorrhoea in the first or second cycle after commencing griseofulvin therapy. Pregnancies were reported in 2 women. Pregnancy was also reported in the CSM data (Table 3) and in a recent case report by Cote (1990). Although griseofulvin has been shown to modify hepatic enzyme activity in mice (Denk et al., 1977) there is no good evidence of a major inducing effect in humans.

1.5 Vitamin C

There are few examples of drugs increasing the therapeutic effect of OCS. Theoretically any drugs which inhibit the specific enzymes involved in OC metabolism could lead to an elevation of plasma concentrations. Thus, if we can find out which drugs are metabolized by P450lllA4 and the other P450 isozymes involved in EE_2 2-hydroxylation, it may be possible to predict drug interactions with EE_2. This of course presupposes that by inhibition of one enzyme pathway there is not an increase in turnover of steroid by an alternative pathway.

Vitamin C (ascorbic acid) is occasionally used in high doses and since, like EE, it is metabolised to sulfate conjugates, there is the possibility of transient overload of sulfation in the intestinal wall and/or the liver is the vitamin C is coadministered with a combined OC.

Two small clinical studies in 6 and 5 subjects respectively (Back et al 1980b, Back et al 1981c) seemed to confirm this interaction with EE bioavailability increased by about 50% when 1g of Vitamin C was given half hour before pill administration. However, the variability of the results was considerable and a much larger group of subjects was necessary to adequately define treatment-related differences. A recent study (Zamah et al 1993) has reinvestigated the interaction in 37 women using a combined OC for two consecutive cycles. Concomitant daily administration of 1g Vitamin C taken half hour before pill intake was randomly assigned to the first or second cycle of OC usage. No effect of Vitamin C was observed on EE AUC_{0-12h} on day 1 or day 15. There was a significantly lower AUC for EE sulfate in the presence of Vitamin C on day 15 but the decline in sulfation was insufficient to alter plasma EE concentrations and is therefore unlikely to be of any clinical importance.

1.6 Paracetamol

Paracetamol is metabolized primarily by conjugation with sulphate and glucuronic acid when given in therapeutic doses in vivo, but some microsomal oxidation leading to the formation of cysteine and mercapturate conjugates also occurs (Prescott, 1980). Since single dose paracetamol had been shown to cause partial depletion of inorganic sulphate in man (Levy et al., 1982; Hendrix-Treacey et al., 1986), Rogers et al. (1987a,b) conducted both in vitro and in vivo studies on the potential for paracetamol to interfere with EE_2 sulphation in the gut wall.

Using a technique which involves isolating mucosal sheets from histologically normal ileum and mounting them between two chambers (Ussing Chambers) it was shown that paracetamol caused a significant decrease in EE_2 sulphation (Rogers et al., 1987a). In the control chambers 33% of EE_2 was sulphated, whereas in the presence of paracetamol (1 mg) this was reduced to 7%. The observations indicated that competition for sulphation was occurring when both drugs were presented to the intestinal mucosa. It was suggested that a decrease in PAPS (3'-phosphoadenosine-5'-phosphosulphate) pool is responsible for the decrease in steroid conjugation.

This in vitro study was followed up by examining the effect of a conventional single dose of paracetamol (1g) on the plasma concentrations of EE_2 and LNG in healthy volunteers in the 24 h period following ingestion of an OCS (Rogers et al., 1987b). The area under the curve of EE_2 was significantly increased following paracetamol administration with the greatest effect evident in the time period 0-3h. There was a significant decrease in AUC of EE_2-sulphate after paracetamol (7736 $\pm$ 3791 pg.ml^{-1}.h) compared to the control (13161 $\pm$ 4535 pg.ml^{-1}.h; $P < 0.05$). The authors concluded that the increase in plasma EE_2 concentrations resulted from a reduction in sulphation of the steroid, substantially in the gut wall, and that this interaction could be of clinical significance in women on OCS who regularly take paracetamol.

1.7 Smoking

Polycyclic hydrocarbons in cigarette smoke are potent inducers of certain cytochrome P450 isozymes (notably P4501A2) and thereby increase the oxidative metabolism of a number of drugs. For example, smoking is associated with enhanced metabolism of antipyrine, propranolol and theophylline (Vestal et al., 1979; Wood et al., 1979; Jusko et al., 1978; Vestal et al., 1987). In contrast, no effect has been observed on the disposition of other drugs (eg. diazepam, phenytoin) reviewed by Jusko (1979). With respect to steroid metabolism, Michnovicz et al. (1986) have reported a marked increase in estradiol (E_2) 2-hydroxylation in smokers with a consequent decrease in circulating E_2 levels. However, Crawford et al. (1981) examined a total of 311 OC users (102 smokers; 209 non-smokers) and found no significant difference in EE_2 plasma concentration profiles in smokers and non-smokers. The clinical studies point to the involvement of different P450 isozymes in the metabolism of E_2 and EE_2. There is some recent experimental data in support of this from Ball et al. (1990) who showed that E_2 is metabolized by proteins from the P4501A (polycyclic aromatic hydrocarbon inducible) 4501IC and P4501IE gene families, whereas EE_2 is catalysed by cytochromes from the P4501IC, P4501IE and P4501IIA gene families.

Thus proteins from the P4501A family appear to play an important role in E_2 metabolism but not in EE_2 disposition. Therefore, from the in vitro studies we would not expect EE_2 to be affected by smoking, which is what is found in the pharmacokinetic studies. Little is known as yet about progestogen metabolism and smoking from an enzymological perspective.

1.8 Adsorbents

Adsorbents such as magnesium trisilicate, aluminum hydroxide, activated charcoal and kaolin can absorb a wide variety of substances onto their surfaces thereby preventing the absorption of drugs and toxins from the gastrointestinal tract. Although on theoretical grounds adsorbents could be expected to interfere with the efficacy of OCS there is no firm evidence that this is the case. Joshi et al. (1986) investigated the bioavailability of OCS in women using antacids (magnesium trisilicate + aluminum hydroxide). As judged by peak concentrations and AUCs of EE_2 and LNG, antacid administration had no effect on OC absorption.

Magnesium-containing antacid preparations can induce diarrhoea and this raises the question of whether infective or induced diarrhoea will lead to a reduced absorption of OC. Sparrow (1987) and Shenfield (1986) are convinced that diarrhoea can induce failure of OC efficacy, and indeed Sparrow records 23 cases of OC failure associated with diarrhoea. However a major consideration is that absorption takes place high in the small bowel since the bioavailability of OCS is not reduced at all in patients with an ileostomy (Grimmer et al., 1986). We know also of a case report (Back & Orme - unpublished) of a woman who had resection of all but 30 cm of small bowel and yet showed normal bioavailability of both EE_2 and LNG. So while the case for diarrhoea being an important consideration of OC failure is attractive the pharmacokinetic data is as yet lacking.

1.9 Other Drugs

Reports have been published with regard to cases of hepatic cholestasis in women taking both troleandomycin and OCS (Fevery et al., 1983; Haber et al., 1980). Oxidation of troleandomycin by a P450 isozyme is known to produce a derivative (probably a nitroso derivative) which binds tightly to the enzyme and thereby causes inactivation. Guengerich (1988) has indicated that this inhibition is highly selective for P450lllA4 and since this isozyme is important in EE_2 metabolism, hepatic accumulation of EE_2 is possible.

There are in the literature a number of isolated case reports of possible interactions between OCS and a variety of other drugs including antihistamines and antiasthmatic drugs. The evidence for interaction is not convincing either from a clinical or from a mechanistic viewpoint.

2. ORAL CONTRACEPTIVES INTERFERING WITH THE METABOLISM OF OTHER DRUGS

There were early indications from animal studies that OCS could inhibit the metabolism of other compounds undergoing oxidative metabolism (Tephly & Mannering, 1968). More recently Guengerich (1988, 1990) showed that acetylenic steroids can become covalently attached to P450 heme after enzymatic oxidation. He demonstrated that EE_2 and progestogens preincubated with human liver microsomes caused a loss of both spectrally detectable cytochrome P-450 and enzyme activity. These in vitro studies provide some mechanistic basis to the many reports in the literature of an inhibitory effect (albeit normally small) of OCS on a number of concurrently administered drugs.

In addition OCS appears to have the propensity to induce glucuronidation. Hence this will have the opposite pharmacokinetic effect to the inhibitory action on oxidation.

2.1 Benzodiazepines

Oral contraceptive steroids have been shown to influence the disposition of some benzodiazepines undergoing oxidation, eg., chlordiazepoxide (Roberts et al., 1979) and diazepam (Abernethy et al., 1982); nitroreduction, eg., nitrazepam (Jochemsen et al., 1982) and conjugation, eg., temzaepam (Stoehr et al., 1984). For benzodiazepines metabolized by oxidative pathways and by nitroreduction, OCS inhibit enzyme activity and reduce metabolic clearance. For example, the elimination half life of diazepam was significantly longer (69 $\pm$ 9 vs 47 $\pm$ 4h) and the total metabolic clearance significantly less (0.27 $\pm$ 0.02 vs 0.45 $\pm$ 0.04 ml.min^{-1}.kg^{-1}) in OC users compared to control subjects in the study of Abernethy et al. (1982). Inhibition of the intrinsic clearance of nitrazepam was seen in OC users by Jochemsen et al. (1982). In non-OC users clearance was 459 $\pm$ 40 ml.min^{-1} whereas in OC users it was 323 $\pm$ 30 ml.min^{-1}. It should be noted that although impaired drug oxidation has been reported for the drugs outlined above, there are others undergoing oxidative metabolism eg. bromazepam (Ochs et al., 1987), clotiazepam (Ochs et al., 1984), alprazolam (Scarone et al., 1988) which show similar pharmacokinetics in both OC users and non-users. In contrast to the inhibition of clearance of some benzodiazepines undergoing oxidative metabolism or nitroreduction, the clearance of temazepam, which is metabolized by glucuronic acid conjugation is increased by 60% (Stoehr et al., 1984). However, the metabolic clearance of lorazepam, which also undergoes conjugation, remained unaltered in two studies (Stoehr et al., 1984; Abernethy et al., 1983) but increased in another (Patwardhan et al., 1983).

Despite the kinetic changes outlined there is very little evidence that this is an interaction of clinical importance.

2.2 Cyclosporin

A potentially important interaction was published as a case report by Deray et al. (1987). A 32 year old woman was treated with cyclosporin for idiopathic uveitis, her OC therapy having been stopped 2 months previous. Reintroduction of the OC was associated with increases in plasma concentrations of cyclosporin; aminotransferase, serum bilirubin and alkaline phosphotase also increased.

Recent in vitro studies with human liver have provided evidence that cyclosporin undergoes hydroxylation and N-demethylation and that P450lllA4 is involved (Kronbach et al., 1988). Again, this is the same isozyme which is at least partially responsible for EE$_2$ 2-hydroxylation (Guengerich, 1988; Ball et al., 1990).

2.3 Analgesic Drugs

Miners et al. (1986) investigated the influence of OCS on salicylic acid and acetylsalicytic acid disposition. Salicylic acid clearance was 41% greater in OCS-users compared to non-OCS users due to increases in conjugation pathways (glycine and glucuronic acid). Increased glucuronidation has also been shown for paracetamol (Miners et al., 1983; Mitchell et al., 1983) and suggested to be the mechanism whereby OCS enhance morphine clearance (Watson et al., 1986).

2.4 Corticosteroids

There are a number of studies reporting marked changes in prednisolone pharmacokinetics in OCS users (Boekenoogen et al., 1983; Meffin et al., 1984; Frey et al., 1984; Legler & Benet, 1986). These alterations, characterized by decreased clearance and volume of distribution and prolonged elimination half life, result from changes in both protein binding (which is increased due to increased CBG levels) and unbound clearance (which is decreased presumably due to inhibition of metabolism). Legler & Benet (1986) proposed careful monitoring of women taking OCS who were concurrently undergoing prednisolone therapy. They suggested that lower doses of prednisolone should yield clinical efficacy in such subjects.

2.5 Ethanol

The possibility of a pharmacokinetic interaction between OCS and ethanol was raised by Jones and Jones (1984) who found a significantly decreased ethanol elimination rate in women taking OCS compared to non OC-users. The implications of this work were that alcohol would be present longer in women taking OCS. However, in another study (Hobbes et al., 1985), no significant effect of OCS on mean peak plasma ethanol concentration, mean time to peak, mean AUC or mean rate of ethanol disappearance was seen. One surprising finding in this study was that OCS improved tolerance to alcohol, although the reasons for this were not clear.

2.6 Other Drugs

Other isolated reports of OCS influencing the kinetics of concomitantly administered drugs include: an increase in plasma concentrations of metoprolol (Kendall et al., 1982); an increase in plasma clearance of clofibric acid, a drug largely metabolized by glucuronidation (Miners et al., 1984); a change in the mean metabolic ratio of extensive metabolisers of debrisoquine (Kallio et al., 1988). See Table 5.

3. INTERACTION BETWEEN PROGESTOGEN AND ESTROGEN COMPONENTS OF OC

Recently, it has been reported that women who received a combination of gestodene and EE_2 (75 ug/30ug) showed higher serum levels of EE_2 than women who received a combination of desogestrel and EE_2 (150 ug/30 ug) despite the fact that the EE_2 dose was the same in both preparations (Jung-Hoffman & Kuhl, 1989). The difference in EE_2 levels between the treatment groups was observed even following the first oral administration of the two preparations. The authors attributed the difference to a possible inhibitory effect of gestodene on the hepatic cytochrome P-450 system, thereby giving rise to elevated EE_2 levels. Subsequent studies (Kuhnz et al, 1990; Humpel et al, 1990, Orme et al 1990, Hammerstein et al 1993), although of different design, have failed to find any differences in EE_2 plasma level when the gestodene-containing and desogestrel-containing formulations were used.

Other controversial findings in this area relate to in vitro studies. Guengerich (1990) has claimed that gestodene is a particularly potent mechanism-based inactivator of cytochrome P-450lllA4, ie, it inhibits the 2-hydroxylation of EE_2 (and other drugs) and this may therefore account for the increased EE_2 plasma levels in some women using the

Table 5. Drugs for which there is evidence of altered pharmacokinetics in OC users.

Increased concentrations	Decreased concentrations
Antipyrine	Acetylsalicylic acid
Chlordiazepoxide	Clofibric acid
Diazepam	Morphine
Alprazolam	Paracetamol
Nitrazepam	Temazepam
Theophylline	
Caffeine	
Cyclosporin	
Prednisolone	
Metoprolol	

gestodene-containing OCS. However, Back et al. (1990) have shown that all the progestogens in common use have the propensity to inhibit cytochrome P450 catalyzed oxidation but there is little evidence for one progestogen being more markedly inhibitory than others.

4. CONCLUSIONS

There are pharmacokinetic drug interactions with oral contraceptives of definite clinical relevance (anticonvulsants, rifampicin) because in the majority of subjects who take the two drugs interaction will occur. Data are emerging concerning the cellular/molecular basis for the interactions. This is not so for the interaction of antibiotics and OCS and we still await some method of being able to predict which women may be at risk. Any interaction which causes raised concentrations of EE_2 is of potential importance because of the fear that high blood levels will correlate to adverse events. To date, there is some evidence that troleandomycin inhibits EE_2 2-hydroxylation and paracetamol competes for sulphation. It is unlikely, however, that many women will be using any of the "offending" drugs for long enough periods to exhibit a clinically significant interaction.

Acknowledgements

We gratefully acknowledge the assistance of many colleagues who have worked in our laboratories and performed many of the studies described in this review.

We are grateful to Sue Oliphant for typing the manuscript.

References

1. Abernethy DR, et al. Impairment of diazepam metabolism by low-dose estrogen-containing oral-contraceptive steroids. *New England Journal of Medicine* 306:791-792. 1982.
2. Abernethy DR, et al. Lorazepam and oxazepam kinetics in women on low-dose oral contraceptives. *Clinical Pharmacology and Therapeutics* 33:628-632. 1983.
3. Abernethy DR, Todd EL. Impairment of caffeine clearance by chronic use of low-dose oestrogen-containing oral contraceptives. *European Journal of Clinical Pharmacology* 28:425- 428. 1985.
4. Back DJ, et al. The interaction of phenobarbital and other anticonvulsants with oral contraceptive steroid therapy. *Contraception* 22:495-503. 1980a.
5. Back DJ, *et. al.* Metabolism by gastrointestinal mucosa-clinical aspects, in Prescott and Nimmo (Eds) *Drug Absorption.* pp 80-87 (ADIS Press, Sydney. 1980b).
6. Back DJ, *et. al.* The in vitro metabolism of ethinyloestradiol, mestranol and levonorgestrel by human jejunal mucosa. *British Journal of Clinical Pharmacology* 11:275-278. 1981b.
7. Back DJ, *et. al.* The effect of rifampicin on the pharmacokinetics of ethinyloestradiol in women. *Contraception* 21:135-143. 1980c.
8. Back DJ, *et. al.* An investigation of the pharmacokinetics of ethinylestradiol in women using radioimmunoassay. *Contraception* 20:263-273. 1979a.
9. Back DJ, *et. al.* The effect of rifampicin on norethisterone pharmacokinetics. *European Journal of Clinical Pharmacology* 15:193-197. 1979b.
10. Back DJ, *et. al.* Interindividual variation and drug interactions with hormonal steroid contraceptives. *Drugs* 21:46-61. 1981a.
11. Back DJ, *et. al.* An antibiotic interaction with ethinyloestradiol in the rat and rabbit. *Journal of Steroid Biochemistry* 16:407-413. 1982a.
12. Back DJ, *et. al.* The interaction of ethinyloestradiol with ascorbic acid in man. *British Medical Journal* 282:1516. 1981c.
13. Back DJ, *et. al.* The gut wall metabolism of ethinyloestradiol and its contribution to the presystemic metabolism of ethinyloestradiol in humans. *British Journal of Clinical Pharmacology* 13:325-330. 1982b.
14. Back DJ, *et. al.* The effects of ampicillin on oral contraceptive steroids in women. *British Journal of Clinical Pharmacology* 14:43-48. 1982c.

15. Back DJ, Breckenridge AM. Orme ML'E. Drug interactions with oral contraceptives. *International Planned Parenthood Federation* 17:1-2. 1983.

16. Back DJ, *et. al.* Evaluation of Committee on Safety of Medicines yellow card reports on oral contraceptive drug interactions with anticonvulsants and antibiotics. *British Journal of Clinical Pharmacology* 25:527-532. 1988.

17. Back DJ, Orme ML'E. Pharmacokinetic drug interactions with oral contraceptives. *Clinical Pharmacokinetics* 18:472-484. 1990.

18. Back DJ, *et. al.* Effect of progestogens on the metabolism of ethinyloestradiol by human liver microsomes in vitro. *British Journal of Clinical Pharmacology* 30:321. 1990.

19. Bacon JF, Shenfield GM. Pregnancy attributable to interaction between tetracycline and oral contraceptives. *British Medical Journal* 1:293. 1980.

20. Bainton R. Interaction between antibiotic therapy and contraceptive medication. *Oral Surgery* 61:453-455. 1986.

21. Ball SE, *et. al.* Differences in the cytochrome P450 isozymes involved in the 2-hydroxylation of estradiol and 17α-ethinyloestradiol: relative activities of rat and human liver enzymes. *Biochemical Journal* 167:221-226. 1990.

22. Belaisch J, Driguez P, Janaud A. Influence de certains medicaments sur l'action des pilules contraceptifs. *Nouvelle Presse Medicale* 5:1645-1646. 1976.

23. Boehm FH, Di Pietro DL, Gross DA. The effect of ampicillin administration on urinary estriol and serum estradiol in the normal pregnant patient. *American Journal of Obstetrics and Gynaecology* 119:98-103. 1974.

24. Boekenoogen SJ, Szefler SJ, Jusko WJ. Prednisolone disposition and protein binding in oral contraceptive users. *Journal of Clinical Endocrinology and Metabolism* 56:701-709. 1983.

25. Chapman CR. Absorption and metabolism of steroid prodrugs. Ph.D. Thesis, University of Liverpool 1981.

26. Combalbert, *et. al.* Metabolism of cyclosporin A. IV. Purification and identification of the rifampicin-inducible human liver cytochrome P450 (Cyclosporin A oxidase) as a product of P450IIIA Gene Subfamily. *Drug Metabolism and Disposition* 17:197-207. 1989.

27. Coté J. Interaction of griseofulvin and oral contraceptives. Journal of the American Academy of *Dermatology* 22:124-125. 1990.

28. Coulam CB, Annegers JF. Do anticonvulsants reduce the efficacy of oral contraceptives. *Epilepsia* 20:519-526. 1979.

29. Crawford FE, *et. al.* Oral contraceptive steroid plasma concentrations in smokers and non-smokers. British *Medical Journal* 182:1829-1830. 1981.

30. Crawford P, *et. al.* The lack of effect of sodium valproate on the pharmacokinetics of oral contraceptive steroids. *Contraception* 33:51-59. 1986.

31. Crawford P, *et. al.* The interaction of phenytoin and carbamazepine with combined oral contraceptives. *British Journal of Clinical Pharmacology* 30:892-896. 1990.

32. Denk H, *et. al.* Alteration of hepatic microsomal enzymes by griseofulvin treatment in mice. *Biochemical Pharmacology* 16:1125-1130. 1977.

33. Deray G, *et. al.* Oral contraceptive interaction with cyclosporin. *Lancet* 1:158-159. 1987.

34. De Sano EA, Hurley SC. Possible interactions of antihistamines and antibiotics with oral contraceptive effectiveness. *Fertility and Sterility* 37:853-854. 1982.

35. Diamond MP, *et. al.* Interaction of anticonvulsants and oral contraceptives in epileptic adolescents. *Contraception* 31:623-632. 1985.

36. Dossetor J. Drug interactions with oral contraceptives. *British Medical Journal* 4:467-468. 1975.

37. Fevery J, Van Steenbergen W, Desmet V. Severe intrahepatic cholestasis due to the combined intake of oral contraceptives and triacetyloleandomycin. *Acta Clinica Belgique* 38:242-245, 1983.

38. Fraser IS, Jansen RPS. Why do inadvertent pregnancies occur in oral contraceptive users. *Contraception* 27:531-551. 1983.

39. Frey BM, Schard HJ, Frey FJ. Pharmacokinetic interaction of contraceptive steroids with prednisone and prednisolone. *European Journal of Clinical Pharmacology* 26:505-511. 1984.

40. Friedman CI, *et. al.* The effect of ampicillin on oral contraceptive effectiveness. *Obstetrics and Gynecology* 55:33-37. 1980.

41. Gagnaire JC, *et. al.* Grossesses sous contraceptifs oraux chex les patients recevant des barbituriques. *Nouvelle presse Medicale* 4:3008. 1975.

42. Grimmer SFM, *et. al.* The effect of cotrimoxazole on oral contraceptive steroids in women. *Contraception* 28:53-39. 1983.

43. Grimmer SFM, *et. al.* The bioavailability of ethinyloestradiol and levonorgestrel in patients with an ileostomy. *Contraception* 33:51-59. 1986.

44. Guengerich FP. Oxidation of 17α-ethynylestradiol by human liver cytochrome P-450. *Molecular Pharmacology* 33:500-508. 1988.

45. Guengerich FP. Mechanism-based inactivation of human liver microsomal cytochrome P450ll-1A4 by gestodene. *Chemical Research in Toxicology* 3:363-371. 1990.

46. Haber I, Hubens H. Cholestatic jaundice after triacetyloleandomycin and oral contraceptives. *Acta Gastroenterologica Belgique* 43:475-482. 1980.

47. Hammerstein J, et al. Influence of gestodene and desogestrel as components of low-dose oral contaceptives on the pharmacokinetics of ethinyl estradiol (EE$_2$) on serum CBG and on urinary cortisol and 6β-hydroxycortisol. *Contraception* 47:263-281. 1993.

48. Helton ED, Williams MC, Goldzieher JW. Human urinary and liver conjugates of 17$_\alpha$-ethynyloestradiol. *Steroids* 27:851-867. 1976.

49. Hempel Von E, *et. al.* Meditamentose enzyminduktion und hormonale kontrazeption. *Zentrablatt fur Gynakologie* 95:1451-1457. 1973.

50. Hendrix-Treacey S, *et. al.* The effect of acetaminophen administration on its disposition and body stores of sulphate. *European Journal of Clinical Pharmacology* 30:273-278. 1986.

51. Hobbes J, Boutagy J, Shenfield GM. Interactions between ethanol and oral contraceptive steroids. *Clinical Pharmacology and Therapeutics* 38:371-380. 1985.

52. Humpel M, et al. Comparison of serum ethinylestradiol, sex-hormone binding globulin, corticoid-binding globulin and cortisol levels in women using two low-dose combined oral contraceptives. *Hormone Research* 33:35-39. 1990.

53. Janz D, Schmidt D. Antiepileptic drugs and failure of oral contraceptives. *Lancet* 1:113. 1974.

54. Jochemsen R, *et. al.* Influence of sex, menstrual cycle and oral contraception on the disposition of nitrazepam. *British Journal of Clinical Pharmacology* 13:319-324. 1982.

55. Jones MK, Jones BM. Ethanol metabolism in women taking oral contraceptives. *Alcoholism: Clinical and Experimental Research* 8:24-28. 1984.

56. Joshi JV, *et. al.* A study of interaction of low dose combination oral contraceptive with ampicillin and metronidazole. *Contraception* 22:643-652. 1980.

57. Joshi JV, *et. al.* Antacid does not reduce the bioavailability of oral contraceptive steroids in women. *International Journal of Clinical Pharmacology, Therapeutics and Toxicology* 24:192-195. 1986.

58. Jung-Hoffmann C, Kuhl H. Interaction with the pharmacokinetics of ehtinylestradiol and progestogens contained in oral contraceptives. *Contraception* 40:299-312. 1989.

59. Jusko WJ. Influence of cigarette smoking on drug metabolism in man. *Drug Metabolism Reviews* 9:221-236. 1979.

60. Jusko WJ, *et. al.* Enhanced biotransformation of theophylline in marijuana and tobacco smokers. *Clinical Pharmacology and Therapeutics* 24:406-410. 1978.

61. Kallio J, *et. al.* Debrisoquine oxidation in a Finnish population: the effect of oral contraceptives on the metabolic ratio. *British Journal of Clinical Pharmacology* 26:791-795. 1988.

62. Kanarkowski R, *et. al.* Pharmacokinetics of single and multiple doses of ethinylestradiol and levonorgestrel in relation to smoking. *Clinical Pharmacology and Therapeutics* 43:23-31. 1988.

63. Kendall MJ, *et. al.* Metoprolol pharmacokinetics and the oral contraceptive pill. *British Journal of Clinical Pharmacology* 14:120-122. 1982.

64. Kenyon TE. Unplanned pregnancy in an epileptic. *British Medical Journal* i:686-687. 1972.

65. Kronbach T, Fischer V, Meyer UA. Cyclosporine metabolism in human liver: Identification of a cytochrome P450lll gene family as the major cyclosporine-metabolizing enzyme

explains interactions of cyclosporine with other drugs. *Clinical Pharmacology and Therapeutics* 43:630-635. 1988.

66.   Kuhnz W et al. Relative bioavailability of ethinyl estradiol from two different oral contraceptive formulations after single oral administration to 18 women in an intraindividual cross-over design. *Hormone Research* 33:40-44. 1990.

67.   Legler UF, Benet LZ. Marked alterations in dose-dependent prednisolone kinetics in women taking oral contraceptives. *Clinical Pharmacology and Therapeutics* 39:425-429. 1986.

68.   Leigh DA, *et. al.* Talampicillin: A new derivative of ampicillin. *British Medical Journal* 2: 1378-1380. 1976.

69.   Levy G, Galinsky R, Lin JH. Pharmacokinetic consequences and toxicologic implications of endogenous cosubstrate depletion. *Drug Metabolism Reviews* 13:1009-1020. 1982.

70.   Lumley CE, *et. al.* The under-reporting of adverse drug reactions seen in General Practice. *Pharmaceutical Medicine* 1:205-212. 1986.

71.   Maggs JL, *et. al.* The biliary and urinary metabolites of ^{3}H-17$_\alpha$-ethinyloestradiol in women. *Xenobiotica* 13:421-431. 1983.

72.   Meffin PJ, *et. al.* Alterations in prednisolone disposition as a result of oral contraceptive use and dose. *British Journal of Clinical Pharmacology* 17:655-664. 1984.

73.   Michnovicz JJ, *et. al.* Increased 2-hydroxylation of estradiol as a possible mechanism for the anti-estrogenic effect of cigarette smoking. *New England Journal of Medicine* 315:1305-1309. 1986.

74.   Miners JO, Attwood J, Birkett DJ. Influence of sex and oral contraceptive steroids on paracetamol metabolism. *British Journal of Clinical Pharmacology* 16:503-509. 1983.

75.   Miners JO, *et. al.* Influence of gender and oral contraceptive steroids on the metabolism of salicylic acid and acetylsalicylic acid. *British Journal of Clinical Pharmacology* 22:135-142. 1986.

76.   Miners JO, Robson RA, Birkett DJ. Gender and oral contraceptive steroids as determinants of drug glucuronidation: effects on clofibric acid elimination. *British Journal of Clinical Pharmacology* 18:240-243. 1984.

77.   Mitchell MC, *et. al.* Effects of oral contraceptive steroids on acetaminophen metabolism and elimination. *Clinical Pharmacology and Therapeutics* 34:48-53. 1983.

78.   Ochs HR, *et. al.* Bromazepam pharmacokinetics: Influence of age, gender, oral contraceptives, cimetidine and propranolol. *Clinical Pharmacology and Therapeutics* 41:562-570. 1987.

79.   Ochs HR, *et. al.* Disposition of Clotiazepam: Influence of age, sex, oral contraceptives, cimetidine, isoniazid and ethanol. *European Journal of Clinical Pharmacology* 26:55-59. 1984.

80.   Orme ML'E. The clinical pharmacology of oral contraceptive steroids. British Journal of Clinical *Pharmacology* 14:31-42. 1982.

81.   Orme ML'E, Back DJ. Drug interactions with oral contraceptive steroids. *Pharmacy International* 1:38-41. 1980.

82.   Orme ML'E, Back DJ. Interactions between oral contraceptive steroids and broad spectrum antibiotics. *Clinical and Experimental Dermatology* 11:327-331. 1986.

83.   Orme ML'E, Back DJ, Breckenridge AM. Clinical pharmacokinetics of oral contraceptive steroids. *Clinical Pharmacokinetics* 8:95-136. 1983.

84.   Orme M'LE et al. The pharmacokinetics of ethinyl estradio in the presence and absence of gestodene and desogestrel. *Contraception* 43:305-316. 1991.

85.   Patwardhan RV, *et. al.* Differential effects of oral contraceptive steroids on the metabolism of benzodiazepines. *Hepatology* 3:248-253. 1983.

86.   Prescott LF. Kinetics and metabolism of paracetamol and phenacetin. *British Journal of Clinical Pharmacology* 10:291S-298S. 1980.

87.   Reimers D, Jezek A. Rifampicin und andere antituberkulostatika bei gleichzeitiger oraler kontrazeption. *Praxis de Pneumologie* 25:255-262. 1971.

88.   Rietvald EC, *et. al.* Rapid onset of an increase in caffeine residence time in young women due to oral contraceptive steroids. *European Journal of Clinical Pharmacology* 26:371-373. 1984.

89.   Roberton YR, Johnson ES. Interactions between oral contraceptives and other drugs. A review. *Current Medical Research and Opinion* 3:647-661. 1976.

90. Roberts RK, *et. al.* Disposition of chlordiazepoxide: Sex differences and effects of oral contraceptives. *Clinical Pharmacology and Therapeutics* 25:826-831. 1979.

91. Roberts RK, *et. al.* Oral contraceptive steroids impair the elimination of theophylline. *Journal of Laboratory and Clinical Medicine* 101:821-825. 1983.

92. Robson RA, *et. al.* Characterisation of theophylline metabolism in human liver microsomes. *British Journal of Clinical Pharmacology* 24:293-300. 1987.

93. Rogers SM, Back DJ, Orme ML'E. Intestinal metabolism of ethinyloestradiol and paracetamol in vitro: studies using Ussing Chambers. *British Journal of Clinical Pharmacology* 23:727-734. 1987a.

94. Rogers SM, *et. al.* Paracetamol interaction with oral contraceptive steroids: increased plasma concentrations of ethinyloestradiol. *British Journal of Clinical Pharmacology* 23:721-725. 1987b.

95. Sahlberg B-L, *et. al.* Analysis of isomeric ethynylestradiol glucuronides in urine. *Journal of Chromatography* 217:453-461. 1981.

96. Shenfield GM. Drug interactions with oral contraceptive preparations. *The Medical Journal of Australia* 144:205-211. 1986.

97. Sparrow MJ. Pill method failures. *New Zealand Medical Journal* 100:102-105. 1987.

98. Stockley I. Drug interactions: An appraisal of the current situation. *Trends in Pharmacological Sciences* 1:6. 1979.

99. Stoehr GP, *et. al.* Effect of oral contraceptives on triazolam, temazepam, alprazolam and lorazepam kinetics. *Clinical Pharmacology and Therapeutics* 36:683-690. 1984.

100. Szoka PR, Edgren RA. Drug interactions with oral contraceptives: Compilation and analysis of an adverse experience report database. *Fertility and Sterility* 49:31-38S. 1988.

101. Tephly TR, Mannering GT. Inhibition of drug metabolism. V. Inhibition of drug metabolism by steroids. *Molecular Pharmacology* 4:10-14. 1968.

102. Van Dijke CPH, Weber JCP. Interaction between oral contraceptives and griseofulvin. *British Medical Journal* 288:1125-1126. 1984.

103. Vestal RE, *et. al.* Aging and drug interactions. I. Effect of cimetidine and smoking on the oxidation of theophylline and cortisol in healthy men. *The Journal of Pharmacology and Experimental Therapeutics* 241:488-500. 1987.

104. Vestal RE, *et. al.* Effects of age and cigarette smoking on the disposition of propranolol in man. *Clinical Pharmacology and Therapeutics* 26:8-15. 1979.

105. Watson KJR, *et. al.* The oral contraceptive pill increases morphine clearance but does not increase hepatic blood flow. *Gastroenterology* 90:1779. 1986.

106. Wood AJJ, *et. al.* Effects of aging and cigarette smoking on antipyrine and indocyanine green elimination. *Clinical Pharmacology and Therapeutics* 26:16-20. 1979.

107. Zamah NM et al. Absence of an effect of high vitamin C dosage on the systemic availbaility of ethinyl estradiol in women using a combination oral contraceptive. *Contraception* 48:377-391. 1993.

A CRITICAL EVALUATION OF THE PHARMACOKINETICS OF CONTRACEPTIVE STEROIDS

Kenneth Fotherby [1]

1. INTRODUCTION

A large number of pharmacokinetic studies of the contraceptive steroids have been performed during the past 25 years but in spite of this our knowledge is still meager. Most of the early studies were performed before many of the problems encountered in such studies were fully realized. The serum concentrations attained after administration of the contraceptive steroids, and hence the pharmacokinetic parameters derived from such measurements, are affected by many different factors. These may have programmatic implications in respect of the acceptability, side effects and beneficial influences of the contraceptive steroids.

This communication will review the different factors which will determine the serum concentration of a contraceptive steroid and the problems encountered in deriving pharmacokinetic parameters, with particular emphasis on variability. Space and time limitations allow only a brief review of each of these areas; more details are available in recent articles[1,2,3].

2. FACTORS AFFECTING SERUM CONCENTRATIONS

2.1 Dose

K. FOTHERBY • Royal Postgraduate Medical School, University of London, London W12 0NN, United Kingdom

Obviously increasing the amount of steroid administered should increase the serum concentration but the relationship between dose and concentration may not be a direct one, e.g. increasing the dose of RU486 from 25mg to 600mg resulted in only a four-fold increase in its serum concentrations[4]. Increasing the dose may reduce the rate of absorption and this will lead to an increase in T_{max} [the time required to reach the peak concentration (C_{max})] and a lower than expected C_{max}. The way in which the dose is presented, i.e. its pharmaceutical formulation, will also be important. Two formulations of equal doses of the same drug, for example, levonorgestrel, ostensibly prepared in the same way, may have different absorption characteristics and hence have different values for T_{max} and C_{max}[5]. These differences were partly due to differences in particle size (micronisation) of the active drug, the smaller the particle size the more efficient the absorption. Such an effect has been described with other steroids[6,7]. The excipients used in the formulation will also have an effect since the faster the rate of dissolution the more rapid will be the absorption.

When the steroids are administered by other routes, e.g. by injection or in a vaginal ring, other factors will be important in determining their uptake. For injectable contraceptives the factors which influence their uptake and hence their serum concentrations and duration of action have been considered in detail previously[8].

All of these factors will introduce variability into the serum concentrations and this will be accentuated when, as with the combined oral contraceptives, two steroids are administered together since they may be affected differently.

2.2 Physiological Factors

The serum concentrations achieved after administration of a given dose of steroid will also be affected by a number of physiological factors. Variations will occur in absorption and also in the enterohepatic circulation which will particularly affect the estrogens. The variability is likely to be increased in subjects who have disorders or infections of the gastro-intestinal tract. For those steroids which undergo a first pass effect this will also determine the amount of the dose reaching the systemic circulation in a biologically active form. In healthy subjects the first pass effect for EE (ethynyloestradiol) varies from 20% to 65% and that of NET (norethisterone) from 47% to 73%. These limits are wide and based on very few estimates. No information on variability in the enterohepatic circulation is available.

In serum the contraceptive steroids are bound to albumin, SHBG, and possibly other proteins and only a small amount is unbound; this latter represents the biologically active fraction. Whereas steroid bound to albumin is freely available, binding to SHBG is much tighter and the factors which control the association and dissociation with SHBG are not clear. EE after administration is rapidly converted to the sulphate and although both the free steroid and the sulphate bind to albumin, neither binds to SHBG and the level of free EE in blood is probably regulated by hydrolysis of the sulphate[9].

The gestagens bind to both albumin and SHBG and the amounts bound to the two proteins depend on the concentration and structure of the gestagen. Estimates of the degree of binding of the various gestagens when administered in oral contraceptives[1] depend on the methodology used. It should be emphasized that the values available are based on very few estimates so that the range of variability is unknown and the extent to which they reflect the *in vivo* situation is questionable. Implicit in all the studies on binding is the assumption that SHBG concentrations are relatively constant between individuals. However in non-treated women there is at least a four-fold variation in the 95% limits for serum SHBG[10]. The possibility exists that there may be inter-population differences in SHBG. Although some data suggest no major differences between British, Chinese, Japanese and Egyptian women in serum SHBG concentrations, the binding of oestradiol to SHBG does appear to differ.

Most estrogens stimulate SHBG synthesis and this can be suppressed by gestagens. The resulting effect depends on the relative doses of oestrogen and gestagen and the structure of the gestagen. There will be a variation among subjects in the resultant response to the oestrogen-gestagen combination. The magnitude of the effect of SHBG on serum gestagen concentration is shown by comparing serum levonorgestrel (LNG) concentrations in women using a combined oral contraceptive containing EE and LNG with those when the subjects are given the same dose of LNG without EE[2]. Serum LNG levels in women receiving LNG with EE are three to four times higher than when receiving LNG alone. The difference has usually been ascribed to LNG binding to the increased amount of SHBG in blood although this explanation has been questioned[2]. The variable response in serum SHBG concentrations to administration of contraceptive steroids will markedly affect the serum steroid levels and hence the pharmacokinetic parameters derived from them.

3. SUBJECT CHARACTERISTICS

3.1 Age and Body Weight

Over the narrow age range of contracepting women, age-related physiological changes are likely to be small. Of more importance are changes in body weight or degree of obesity. The pharmacokinetics of many drugs, particularly lipid-soluble ones, are altered in obesity and there is evidence that the metabolism of some endogenous steroids differs between thin and obese subjects[11]. Some of these changes probably result from an increase in the volume of distribution of the lipid-soluble steroids in the obese subject. In a clinical study of the injectable contraceptive norethisterone oenanthate, the body weight of the women who became pregnant was significantly lower than that of those who did not.

However a study of injectable contraceptives, which eliminated effects due to intestinal absorption, administered to obese women (body mass index >28) and to thin women (BMI <20) did not detect any differences in the pharmacokinetics of the gestagens. A number of reports suggest that dietary lipids, and particularly the blood levels of free fatty acids[12] may affect the binding of endogenous sex hormones to serum proteins making the biologically active fraction more available to the tissues. This finding may be of particular importance in view of the profound effect of the contraceptive steroids on lipid metabolism. SHBG concentrations are associated with body weight, being low in obese subjects and increased in thin ones[10]. Although the evidence suggests that changes in body weight may produce profound changes in the pharmacokinetic of the contraceptive steroids, no conclusive data are yet available.

3.2 Diet

Although food intake affects the absorption of many drugs, information regarding the sex steroids is scarce. The elimination half-life (Tel) for both norethisterone (NET) and LNG was lower in Indian women from a low socio-economic group than in well-nourished women[13]. The mechanism of this effect is not clear and the results conflict with two other findings. Firstly the well-nourished women might have been expected to have a high protein diet and yet high protein diets are claimed to stimulate the hepatic drug metaboliz-ing system leading to a decrease in Tel. Secondly the women in the low socio-economic group were probably mainly on a vegetarian diet and this also has to be taken into account. Diet also affects serum SHBG levels[10] which are increased in subjects on a low fat diet and in vegetarians, and in absence of other changes this should also lead to an increase in Tel, although in one preliminary study there were no significant differences in the pharmacoki-netics of NET between vegetarians and non-vegetarians[14]. High protein diets appear to enhance the activity of the oestrogen 2-hydroxylase whereas high carbohydrate diets appear to decrease its activity; these changes may also be accompanied by changes in 16α-hydroxylase activity.

3.3 Smoking

Smoking affects the incidence of many pathological disorders[15,16]. It has a variable effect on drug metabolism and will enhance that of many drugs[17]. Some steroids appear to be affected but not others[18]. Oestrogen metabolism seems to be particularly influenced; the serum concentrations of administered estrogens can be reduced by smoking[19] and the 2-hydroxylation of oestradiol is increased[20]. These changes appear to result from a stimulation of oestrogen metabolism in the liver, presumably by increasing the activity of the hepatic mixed function oxidase which is known to be stimulated by the polycyclic aromatic hydrocarbons produced as a result of smoking. For the contraceptive steroids, the results, which must be regarded as very preliminary, are equivocal. The rate of metabolism of EE and LNG were not changed by smoking in one study[21] whereas in another[22] variable results were obtained.

3.4 Drug-Steroid Interactions

Concomitant use of a number of drugs will affect the metabolism of the contraceptive steroids. This topic is considered in another presentation at this Symposium and in a recent article[3]. Interactions to an extent that produce clinical effects occur in only a small minority (<5%) of women using contraceptive steroids and occur mainly with anticonvulsants, antibiotics and antibacterial agents. It seems likely however that alterations in metabolism of the contraceptive steroids which do not produce clinical effects occur in many women using the drugs. The metabolism of the steroids may also be changed in some pathological conditions.

3.5 Genetic and Other Factors

Genetic factors are usually regarded as the main determinant of intersubject variations[23] although their importance in respect of the contraceptive steroids is unknown. These factors are mentioned further below.

4. PHARMACOKINETIC PARAMETERS

A number of different approaches have been described for deriving pharmacokinetic parameters from the serum concentrations of the steroid. The parameters will thus be influenced by the factors described above and also by a number of technical considerations.

Firstly, the accuracy of the measurement of the serum concentrations will depend on the reliability criteria of the methodology used. This will be particularly important when serum concentrations are low, for example for EE, where peak concentrations are usually below 200 pg/ml, and for other steroids 24h or more after dosing. The methodology used should be sensitive enough to measure these low concentrations and also be highly specific. Otherwise inaccurate values will be obtained for many of the pharmacokinetic parameters.

Secondly, the number of samples taken for analysis and the timing of the samples will be important. For example if an insufficient number of samples are taken after dosing up to the time of peak concentration, calculation of the absorption rate constant will be difficult, whereas if an insufficient number of samples are taken 8h and more after dosing, determination of Tel and AUC will be inaccurate for those steroids with a long Tel.

Thirdly, most studies have calculated the parameters after a single dose of the steroid and these values may be very different from those obtained under steady state conditions[1]. These latter values will be more appropriate to the clinical use of the formulations.

Although these points may seem obvious, they have not adequately been taken into account in many studies.

Fourthly, problems may arise in deriving the pharmacokinetic parameters from the serum concentrations. The various methods that may be used and their characteristics have been outlined[2] and the values obtained for the parameters will depend on which procedure is used. For the contraceptive steroids the two most often used have been the simple descriptive and compartmental modelling, although both will give only approximate values. The former will enable the parameters most useful to the clinician to be determined from a plot of the serum concentrations against time after dosing. Compartment modelling will allow the derivation of a wider range of parameters, but deciding which model is most appropriate may be difficult. Some variability between the values reported in different studies may be due to the different methods used.

The parameters most useful to the clinician will be T_{max}, C_{max}, Tel and bioavailability as assessed by the area under the serum concentration-time curve (AUC). These are usually regarded as being simple to interpret but this may not always be true. Tel is a secondary rather than primary parameter, being determined by two other parameters, volume of distribution and clearance, and may be a composite of different elimination half-lives. Tel may also depend on whether the gestagen, for example, is administered alone or with an oestrogen (see 2). It may therefore be necessary in quoting a value for Tel to indicate the conditions under which it was obtained. Similarly the AUC for LNG administered with EE may be many times higher than that for the same dose of LNG administered without EE[2]. Thus the bioavailability may appear to be much larger in the former than in the latter situation but this is unlikely to be the case. It should also be emphasized that these determinations of pharmacokinetic bioavailability may bear no relationship to pharmacodynamic bioavailability.

5. VARIABILITY

Intersubject variability has been described for a number of contraceptive steroids whether administered orally or intramuscularly[24] and this has recently been reassessed in conjunction with intrasubject variability for Tel and bioavailability. For Tel of LNG[2] the intersubject variation was 3-fold when administered with EE and 1.9-fold when administered alone compared to intrasubject variability of up to 2.6-fold and up to 1.7-fold respectively. For the AUC of LNG the intersubject variation was 2.2-fold when administered with EE and 2.9-fold when given alone compared to intrasubject variability of up to 1.4-fold and up to 2.4-fold respectively. Thus intrasubject variability for both Tel and AUC for LNG may account for up to almost 90% of intersubject variability.

Two reports have recently provided data on intrasubject variability for EE. In one study[25] the range of extreme values was from -7% to +92% of the mean value. In our study[26] the intersubject variation was up to 3.3-fold for Tel and up to 4-fold for AUC compared to values of up to 2.2-fold and 1.4-fold respectively for AUC. Thus up to 66% of the intersubject variability for Tel and up to 35% for AUC could be accounted for by intrasubject variability.

Equally large variations almost certainly occur in the other pharmacokinetic parameters. These results, which show that variability in an individual in the metabolism of the contraceptive steroids may be significant quantitatively to variation between individuals, brings into question whether genetic factors, which have been claimed to be the more important determinant in the inter-individual variations in metabolism of many drugs, are equally important for the contraceptive steroids.

From the programmatic aspect other types of variability may also be important. Inter-population differences in the metabolism of the contraceptive steroids have been demonstrated[25,27]. Seasonal variations, which may be important in some countries, have not been investigated. Cyclic variations may occur with those contraceptive formulations which do not, or do not always, inhibit ovulation. Diurnal variations occur in the absorption and metabolism of some drugs[28], although a preliminary study comparing morning and evening administration of an oral contraceptive did not show any significant differences[29]. Preliminary evidence suggests that the metabolism of the contraceptive steroids changes on long-term administration.

6. CONCLUSIONS

This brief review has described a number of factors which lead to a wide variation in the serum concentrations of the contraceptive steroids and hence in their pharmacokinetic parameters. Problems arise in both deriving and interpreting these parameters which may lead to confusion. Some of these problems may arise because almost all studies have measured the total concentration of the steroid in blood. More meaningful information might be obtained by estimating for example the free non-protein bound fraction which is regarded as being the biologically active fraction, but current methodology is not adequate on the basis of sensitivity and the wide-scale studies that are necessary. The correlation of pharmacokinetic parameters with pharmacodynamic effects of the contraceptive steroids has not been considered but little evidence is available to support an association. Consideration of all these points suggests that evaluating the pharmacokinetics of the contraceptive steroids has limited usefulness.

References

1.   Fotherby K. Potency and pharmacokinetics of gestagens. *Contraception* 41:533. 1990.
2.   Fotherby K. Pharmacokinetics of gestagens: some problems. *Am J Obstet Gynecol* 163:323.
        1990.
3.   Fotherby K. Interactions with oral contraceptives. *Am J Obstet Gynecol* 163:2153. 1990.
4.   Shi YE, *et. al.* Pharmacokinetic study of RU486. 1990 (In preparation).
5.   He CH, *et. al.* Pharmacokinetic study of two types of post-coital contraceptive containing
        levonorgestrel. *Contraception* 41:557. 1990.
6.   Gibian H, *et. al.* Effect of particle size on biological activity of norethisterone. *Acta Physiol
        Lat Amer* 18:323. 1968.
7.   Saperstein S, *et. al.* Pharmacokinetics of norethindrone; effect of particle size. *Contraception*
        40:731. 1989.
8.   Fotherby K. Factors affecting the duration of action of injectable contraceptives. *Contracept
        Deliv Syst* 2:249. 1981.
9.   Fotherby K. Pharmacokinetics of ethynyloestradiol in humans. *Methods Find Exp Clin
        Pharmacol* 4:133. 1982.
10.   Moore JW, Bulbrook RD. Epidemiology and function of SHBG. *Oxford Reviews of Reproduc-
        tive Biology* 10:180. 1988.
11.   Herschcopf RJ, Bradlow HL. Obesity, diet and estrogens. *Am J Clin Nutr* 45:283. 1987.
12.   Reed MJ, *et. al.* Oestrogen metabolism in women: the role of dietary lipids. *J Ster Biochem*
        27:985. 1987.
13.   Madhavan NK, *et. al.* Pharmacokinetics of LNG in Indian women. *Contraception* 20:303.
        1979.
14.   Prasad RNV, *et. al.* Single dose kinetics of norethisterone in lactovegetarians. *Singapore J
        Obstet Gynaecol* 12:59. 1981.
15.   Fielding JE. Smoking and health effects. *N Eng J Med* 313:491. 1985.
16.   Stillman RJ, Rosenberg MJ, Sachs BJ. Smoking and reproduction. *Fertil Steril* 46:545. 1986.
17.   Hart P, *et. al.* Enhanced drug metabolism in cigarette smokers. *Br Med J* 2:147. 1976.
18.   Friedman AJ, Ravnikar VA, Barbieri, RL. Serum steroid profiles in smokers and non smokers.
        Fertil Steril 47:398. 1987.
19.   Jensen J, Christiansen C, Rodbro, P. Cigarette smoking and serum estrogens during HRT.
        N Engl J Med 313:973. 1985.
20.   Michnovicz JJ, *et. al.* Cigarette smoking alters oestrogen metabolism. *Metabolism* 38:537.
        1989.
21.   Crawford P, *et. al.* Steroid plasma concentrations in smokers and non smokers. *Br Med J*
        282:1829. 1981.
22.   Kanarkowski R, *et. al.* Pharmacokinetics of ethinylestradiol and levonorgestrel in relation to
        smoking. *Clin Pharmacol Ther* 43:23. 1988.
23.   Vessel ES. Genetic and environmental factors affecting drug disposition. *Clin Pharm Ther*
        22:659. 1977.
24.   Fotherby K. Variability of pharmacokinetic parameters of contraceptive steroids. *J Ster
        Biochem* 19:817. 1983.
25.   Goldzieher JW. Pharmacokinetics and metabolism of ethinylestrogens and their clinical impli-
        cations. *Am J Obstet Gynecol* 163:318. 1990.
26.   Fotherby K. Intrasubject variability in the pharmacokinetics of ethynyloestradiol. *J Ster
        Biochem* 38:733. 1991.
27.   Fotherby K, *et. al.* Pharmacokinetics of ethynyloestradiol in women from different populations.
        Contraception 23:487. 1981.
28.   Lemmer B. *Chronopharmacology.* Dekker, New York, 1989.
29.   Kiriwat O, Fotherby K. Pharmacokinetics of oral contraceptives after morning or evening
        administration. *Contraception* 27:153. 1983.

A SHORT REVIEW ON THE SOURCES OF VARIABILITY IN PHARMACOKINETIC PARAMETERS OF SEX STEROIDS: COMMENT ON THE PAPER BY K. FOTHERBY

Michael Hümpel[1]

Four issues were discussed by K. Fotherby in Chapter 11 "A critical evaluation of pharmacokinetics", these were:

*   Factors affecting serum concentrations
*   Subject characteristics
*   Pharmacokinetic parameters and
*   Subject variability

Mention is made of all factors which are either known to affect or are postulated to influence the pharmacokinetics of contraceptive steroids. While agreeing with most of the points raised and conclusions reached there are some points which could be highlighted or given a different weight, these are:

1. *Absorption of dose*

   As far as combination oral contraceptives are concerned the extent and rate of absorption do not seem to contribute to the overall variability. Several studies have shown bioequivalence between various preparations and no cross inter- ference of progestogen and estrogen with respect to absorption[1-4]. Thus variations in absorption should be a very minor part, if any, of the overall variability.

M. HUMPEL • Schering AG, Muellerstrasse 178, 13342 Berlin, Federal Republic of Germany.

2.   *Enterohepatic recirculation and gastro-intestinal tract (g.i.t) disturbances*

Because a direct conjugation of progestogens is of minor quantitative importance in metabolism, enterohepatic recirculation seems to occur with ethinyl estradiol only. However, the scarce information available[5] does not suggest a quantitatively important effect. The secondary increase in plasma ethinyl estradiol levels around 12 hours after a single dose, makes up at the most 10 - 15% of the total area under the curve or has not even been observed at all. Infections of the g.i.t., diarrhea or operative shortening of the small intestine have not been proven to influence sex steroid pharmacokinetics. Such warnings in package insert are not based on specific studies but should be viewed as a general worst case risk estimation.

3.   *Body weight and diet/nutrition*

Admittedly, there is little information on the possible effect of these parameters on sex steroid pharmacokinetics. There was a general agreement that this lack of information could and should be filled by retrospective and prospective studies.

4.   *Dietary lipids and lipid metabolism*

The interrelationship of the effect of transient changes in blood dietary lipids on progestogen protein binding to the observed effect of oral contraceptive use on lipid metabolism detectable after several months of pill intake remains obscure.

5.   *Pharmacokinetic parameters*

Pharmacokinetic parameters of progestogens of the 19-nortestosterone type are significantly influenced by their protein binding when calculated from total drug plasma or serum levels. The binding to SHBG has a particularly great impact because during pill use SHBG levels will or will not be induced by concomitantly administered ethinyl estradiol, the effect being dependent on the preparation used.

Therefore the role of SHBG must be considered when single dose pharmacokinetics and respective parameters calculated during multiple dose studies are compared. For instance, the increase of progestogen plasma levels, a prolongation of the terminal half-life of disposition from plasma and a change in "apparent" total clearance from plasma during a cycle of use of a combination oral contraceptive all can be attributed to the changes in SHBG serum levels[6-9]. These basic interrelationships and their impact on evaluation and interpretation of pharmacokinetics parameters should be considered more.

I do not agree with the statement that the methods used for evaluation of pharmacokinetic parameters themselves contribute to the variability. Obviously, the evaluation of pharmacokinetic studies requires numerous and appropriately timed samples for analysis. Given this, the evaluation of parameters is well defined and not open for individual interpretation. It must be stated that the literature if not free of incomplete sets of data which are pharmacokinetically overinterpreted[6,10-14]. Even the evaluation of a censored set of data[16] leading to discordant results has appeared[15]. Thus,

Table 1. Sources of variability: ADME

	Single Dose	Repeated Dose		
		P only	E only	Comb.OC
Absorption	-	-	-	-
Distribution	-/+	++	-	+++
Metabolism	+++	(-)	(-)	(-)
Excretion	-	-	-	-

P =	Progestin	- =	No increase of variability suggested
E =	Estrogen	+, ++, +++ =	Different degrees of increase in variability
		(-) =	No further increase in variability by repeated treatment suggested

following the accepted, well-defined rules of sampling and evaluation the methods of evaluation of pharmacokinetic parameters will not be a source of variability.

In summary, there are several factors which have or might have an influence on the variability of pharmacokinetics. However, factors such as body weight, age, enterohepatic recirculation, diet, smoking are thought to contribute only a small portion to large overall variability and might have no effect at all on the stability of the pharmacokinetic parameters within a group of sufficient size.

Looking at the processes summarized by the term pharmacokinetics (absorption, distribution, metabolism and excretion), the main source of variability in single dose studies is metabolism (metabolic clearance rate, first-pass metabolism) and distribution (eg. SHBG) in long-term use of progestogens only or of OCs (Table 1).

Following the discussion of possible and proven factors influencing variability in pharmacokinetics the question arises of what the magnitude of the variability is and what its characteristics are. Tables 2 and 3 summarize the coefficients of variation (CV) of the most stable pharmacokinetic parameter, the area under the curve (AUC), for several progestogens and ethinyl estradiol, respectively.

The intersubject CV of various progestogens (Table 2) ranged from 22 to 49% (single dose) and from 22-54% (multiple dose). Although studies cited were of different sizes n = 6 to n = 40) there was no clearcut decrease of the parameter in relation to the studied group size. Therefore it seems that with a group size of 6-12 individuals intersubject CV is high or low by chance (e.g. the CV has itself a large variance). For a stable value a group size of more than 18 individuals seems to be necessary.

Table 2. A short review of the literature on the intersubject coefficient of variation (intersubject CV) in AUC following oral administration of various progestins

A: Single dose

Progestin	NET			LN			KDG			GEST		
	CV(%)	N=	Ref.	CV(%)	N=	Ref.	CV(%)	N=	Ref.	CV(%)	N=	Ref.
	41	24	17	35	20	18	39	9	22	43	11	6
	39	24	17	31	20	18	41	11	10	35	6	23
	44	24	17	39	9	19				47	6	23
	35	24	17	31	9	19				49	6	23
	38	24	1	47	9	20				22	40	24
	35	24	1							28	40	24
Range	35-44			31-47			39-41			22-49		

B. End of treatment cycle (combination oral contraceptive)

	CV(%)	N=	Ref.	CV(%)	N=	Ref.	CV(%)	N=	Ref.	CV(%)	N=	Ref.
				51	18	25	43	11	10	34	11	6
				51	18	25	40	11	10	31	11	6
	No data found			45	8	25	33	11	10	44	11	6
				24	9	22	36	11	10	42	11	6
				22	9	22	54	10	21	36	40	24
										32	40	24
Range				22-51			33-54			31-44		

One would suggest that there are no marked differences in the coefficient of variation of AUC between various progestogens on the basis of the cited references. Intersubject CV should range from 25 to 35% in representative groups.

Evaluation of the same parameter for orally administered ethinyl estradiol (Table 3) revealed that the variability of systemic availability of ethinyl estradiol is not markedly higher than for progestogens. In groups of more than 20 individuals intersubject CV was about 30% after single administration and 30-40% after repeated doses.

In conclusion, the intersubject CV of the most stable pharmacokinetic parameter (AUC) seems to be similar for various progestogens (25-35%) and falls into the same range as is known for other drugs. For ethinyl estradiol a somewhat higher intersubject CV of 30-40% can be expected.

Another question with practical consequence is the intrasubject variability in pharmacokinetics. If, for instance, the systemic availability of ethinyl estradiol showed the same degree of scatter within one individual (day to day) as between individuals, individual dose titration of the estrogen would make no sense. Recently, Goldzieher reported a similarly high intra- and interindividual variation in ethinyl estradiol pharmacokinetics[28]. We analyzed, however, the results of Dibbelt et. al.[25] including ethinyl estradiol levels in 83 women during three months and concluded that the intraindividual variance component in AUC_{0-4h} is about 60-80% of intersubject variance. Scatter plots of AUC_{0-4h} obtained for days 1, 10 and 21 of the first and third cycles gave regression coefficients of 0.54, 0.79 and 0.75, respectively, which is equivalent to an 84%, 61% and 66% intrasubject CV as compared to that between individuals.

A similar evaluation for AUC of the coadministered progestogen-gestodene in 39 women resulted in a 76%, 51% and 51% intrasubject CV as compared to the intersubject CV. Clearly, in the study mentioned the day to day variation within individuals was much lower than among individuals for the two sex steroids. Therefore, attempts to try for an individual dose titration in ethinyl estradiol seem to be justified.

In addition to the scatter of values the distribution has been analyzed using the same parameter (AUC_{0-4h}) for $_2$ and gestodene. AUCs of gestodene (n = 39) and of ethinyl estradiol(n = 83) from the study of Dibbelt et al.[25] clearly showed a Gaussian distribution independent of the day of cycle (days 1, 10, 21) or of treatment cycle (cycles 1 and 3). There was no evidence that the distributions were mixtures of distributions. Therefore, at least in the Caucasian population studied, no evidence was found for genetic polymorphism in sex steroid metabolism or any other genetic predisposition for an enhanced or reduced sex steroid metabolism in subgroups.

The respective analysis of SHBG serum levels revealed an interesting phenomenon. Before OC treatment started serum SHBG levels were distributed normally (n = 83). On day 10 and - even more pronounced - on day 21 of the first treatment

Table 3. A short review of the literature on the intersubject coefficient of variation (CV intersubject) in AUC following oral administration of ethinyl estradiol

| A: Single dose | | | B: Repeated dose (as OC) | | |
CV (%)	N =	Ref.	CV (%)	N =	Ref.
33	18	4	53	18	26
31	18	4	46	18	26
69	11	12	48	8	26
58	11	12	29*	40	25
29	24	1	29*	43	25
28	24	1	40*	11	12
31	9	19	22*	11	12
36	9	19	50*	11	12
28	40	25	41*	11	12
32	43	25	55*	11	12
			37*	11	12
			62*	11	12
			38*	11	12
			39*	30	27
			38*	39	27

* means $AUC_{0-4 h}$, all other values refer to $AUC_{0-24 h}$

cycle with either Femodene (75 ug gestodene + 30 ug ethinyl estradiol) or Marvelon (150 ug desogestrel + 30 ug ethinyl estradiol) bimodal distributions of SHBG levels were found. Independent of the preparation there were two subgroups of women not responding to the ethinyl estradiol dose in the same manner. In one group 30 ug ethinyl estradiol/day induced SHBG serum levels more than in another slowly responding group. This effect turned out to be transient because on days 1, 10 and 21 of the third treatment cycle again a Gaussian distribution was found (Figure 1).

The bimodal distribution of SHBG levels during the first treatment cycle will be evaluated in more detail using demographic, clinical and pharmacokinetic data for correlations. Interestingly CBG levels showed no deviations from the Gaussian distribution at any time of the study.

CONCLUSIONS

1.   Most pharmacokinetic studies published were not designed to serve as one variable for the correlation with clinical side effects. Larger and more specifically planned studies have to be undertaken for such a purpose. However, the hope of a positive outcome of such a possible project is low by both theoretically and from practical experience.

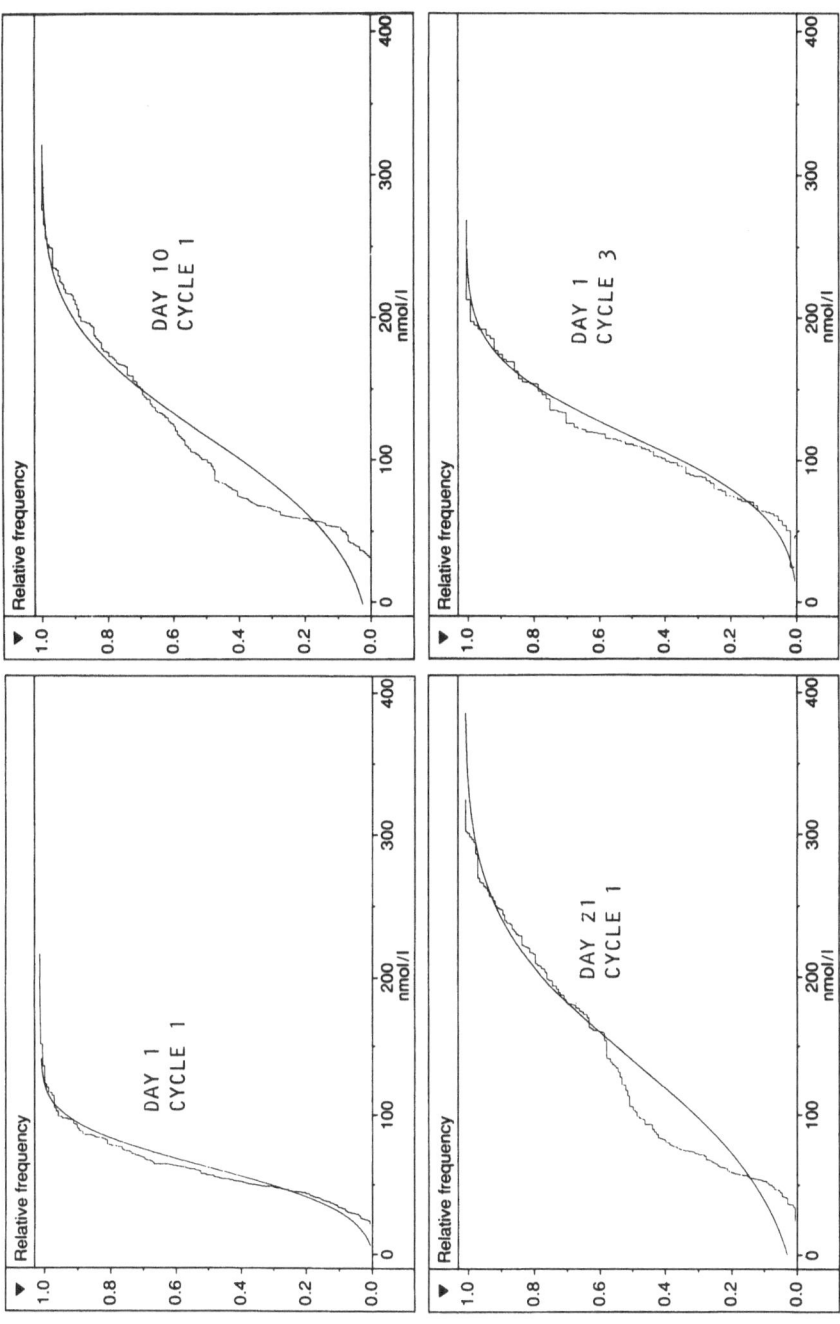

Figure 1. Cumulative frequency distribution of SHBG serum levels from 83 women on days 1, 10 and 21 of the first treatment cycle using one of two low-dose oral contraceptives and on day 1 of the third treatment cycle; preparations used contained 30 μg EE_2 and either 75 μg gestodene (n = 40) or (150 μg desogestrel (n = 43).

2.   Data presented on lower intrasubject CV as compared to intersubject CV still have
to be confirmed, but might give hope for a more optimized   individual   estrogen
treatment in the future.

3.   When appropriate more detailed pharmacokinetic studies assessing the influence of
nutrition and ethnicity should be undertaken in the future. To this end, international
organizations might cooperate with the pharmaceutical industry whenever
appropriate.

Acknowledgement

I gratefully acknowledge that Dr. L. Dibbelt and Dr. D. Back agreed to the use of individual
unpublished data on ethinyl estradiol, SHBG, CBG and gestodene, respectively. All statistical evaluations were carried out by Dr. T. Louton.

References

1.   Saperstein S, *et. al.* Bioequivalence of two oral contraceptive drugs containing norethindrone and
       ethinyl estradiol. *Contraception* 41:581-591. 1989.
2.   Brody SA, Turkes A, Goldzieher JW. Pharmacokinetics of three bioequivalent norethindrone/
       mestranol-50 ug and three norethindrone/ethinyl estradiol-35 ug OC-formulations: Arc
       "low-dose" pills really lower? *Contraception* 40:269-284. 1989.
3.   Fotherby K, Warren RJ. Bioavailability of contraceptive steroids from capsules. *Contraception* 14:
       261-267. 1976.
4.   Kuhnz W, *et. al.* Relative bioavailability of ethinyl estradiol from two different oral contraceptive
       formulations after single oral administration to 18 women in an intraindividual cross-over design.
       Hormone Research 33:40-44. 1990.
5.   Back DJ, *et. al.* An investigation of the pharmacokinetics of ethinylestradiol in women using
       radioimmunoassay. *Contraception* 20:263-273. 1979.
6.   Kuhl H, Jung-Hoffman C, Heidt F. Alterations in the serum levels of gestodene and SHBG during
       12 cycles of treatment with 30 ug ethinylestradiol and 75 ug gestodene. *Contraception* 38:
       477-486. 1988.
7.   Hümpel M, *et. al.* Protein binding of active ingredients and comparison of serum ethinyl estradiol,
       sex hormone-binding globulin, corticosteroid-binding globulin, and cortisol levels in women using
       a combination of gestodene/ethinyl estradiol (Femovan) or a combination of gestodent
       desogestrel/ethinyl estradiol (Marvelon) and single-dose ethinyl estradiol bioequivalence from
       both oral contraceptives. *Am J Obstet Gynecol* 163:329-333. 1990.
8.   Kuhnz W, Pfeffer M, Al-Yacoub G. Protein binding of the contraceptive steroids gestodene, 3-
       ketodesogestrel and ethinylestradiol. *J Steroid Biochem* 35:313-318. 1990.
9.   Smallwood RH, *et. al.* Effect of a protein binding change on unbound and total plasma
       concentrations for drugs of intermediate hepatic extraction. *J Pharmacokin Biopharm* 16:
       529-542. 1988.
10.   Kuhl H, Jung-Hoffman C, Heidt F. Serum levels of 3-ketodesogestrel and SHBG during 12 cycles
       of treatment with 30 ug ethinylestradiol and 150 ug desogestrel. *Contraception* 38:381-390.
       1988.
11.   Goldzieher JW, Dozier TS, de la Pena A. Plasma levels and pharmacokinetics of ethynyl estrogens
       in various populations. *Contraception* 21:1-16. 1980.

12.   Jung-Hoffmann C, Kuhl H. Interactions with the Pharmacokinetics of Ethinylestradiol and Progestogens in Oral Contraceptives. *Contraception* 40:299-312. 1989.
13.   Hümpel M, *et. al.* Investigations of pharmacokinetics of ethinylestradiol to specific consideration of a possible first-pass effect in women. *Contraception* 19:421-429. 1979.
14.   Jenkins N, Limpongsamrak S, Fotherby K. Circulating levels of synthetic steroids in women using a "triphasic" formulation: a comparison with different ethinyloestradiol doses. *J Obstet Gynecol* 2:37-40. 1981.
15.   Fotherby K. Pharmacokinetics of progestational compounds. *Maturitas* 8:123-132. 1986.
16.   Hümpel M, *et. al.* Intraindividual comparison of pharmacokinetic parameters of d-norgestrel, lynestrenol and cyproterone acetate in 6 women. *Contraception* 16:199-215. 1977.
17.   Saperstein S, *et. al.* Pharmacokinetics of Norethindrone. Effect of particle size. *Contraception* 40: 731-740. 1989.
18.   Chang-hai H, *et. al.* Comparative cross-over pharmacokinetic study on two types of postcoital contraceptive tablets containing levonorgestrel. *Contraception* 41:557-567. 1990.
19.   Back D, *et. al.* The relative bioavailability of levonorgestrel and ethinylestradiol when administered in tablet and capsule form. *Contraception* 36:321-326. 1987.
20.   Kuhnz W. Internal report (Microgynon).
21.   Hasenack HG, Bosch AMG, Kaar K. Serum levels of 3-keto-desogestrel after oral administration of desogestrel and 3-ketodesogestrel. *Contraception* 33:591-596. 1986.
22.   Back D, *et. al.* Plasma concentrations of 3-ketodesogestrel after oral administration of desogestrel and intravenous administration of 3-ketodesogestrel. *Contraception* 35:619-626. 1987.
23.   Täuber U, Tack JW, Matthes H. Single dose pharmacokinetics of gestodene in women after intravenous and oral administration. *Contraception* 40:461-472. 1989.
24.   Kuhnz W. Internal data (Schering AG) on Femodene.
25.   Dibbelt L, *et. al.* Group comparison of serum ethinyl estradiol, SHBG and CBG levels in 83 women using two low-dose combination oral contraceptives for three months. *Contraception* (in press)
26.   Kuhnz W. Internal data (Schering AG) on Triquilar.
27.   Hümpel M, *et. al.* Comparison of serum ethinyl estradiol, sex-hormone-binding globulin, corticoid-binding globulin and cortisol levels in women using two low-dose combined oral contraceptives. *Hormone Research* 33:35-39. 1990.
28.   Goldzieher JW. Pharmacology of contraceptive steroids: A brief review. *Am J Obstet Gynecol* 160:1260-1264. 1989.

DIET AND ESTROGEN METABOLISM

Christopher Longcope[1]

The hypothesis that dietary constituents affect estrogen metabolism arose, in part, from epidemiological data suggesting that dietary fat was related to breast cancer[1-3]. These data showed that in populations that ate large quantities of animal fat the incidence of breast cancer was greater than in populations eating minimal amounts of animal fat[1,2,4]. Since there was evidence that estrogens might be associated with breast cancer, it was felt that low fat diets might reduce the risk of breast cancer by lowering estrogen levels.

The initial studies to investigate the possible relationship between diet and estrogens consisted in the measurement of estrogen concentrations in the blood and urine[4]. These studies were usually vertical, i.e., done comparing estrogen concentrations in different population groups, and while many studies showed lower estrogen levels in low fat consumers compared to high fat consumers, there were studies showing little or no difference between these groups[5-9].

Although such studies were non-invasive and could yield important data, they could not delineate between altered production of estrogens and shifts in the pathways of estrogen metabolism without changes in production. These questions might be resolved, however, by studies involving the administration of estrogen tracers. Recently several investigators have used estrogen tracers to measure estrogen production and to characterize the pattern of metabolism in groups of women during perturbations of their diet.

There have been two types of tracer studies that have yielded somewhat different types of information. The first approach involved administration of an estrogen, radio-labeled at a metabolically inert site, and the measurement of radio-labeled parent and products in the blood and urine. From such experiments, one could calculate the clearance rate of the estrogen from the blood and the ratio of the concentrations of radioactivity as product to that of parent in the blood[10]. If one knew the concentration of the endogenous

C. LONGCOPE • Department of OB/GYN, University of Massachusetts Medical School, Worcester, MA 01655

parent estrogen in the blood, one could calculate the blood production rate of that estrogen[10]. By measuring the amounts of radioactivity in specific estrogens excreted in the urine, one could calculate the percentage of administered parent estrogen metabolized to the various products, and thus characterize the pattern of metabolism and relative

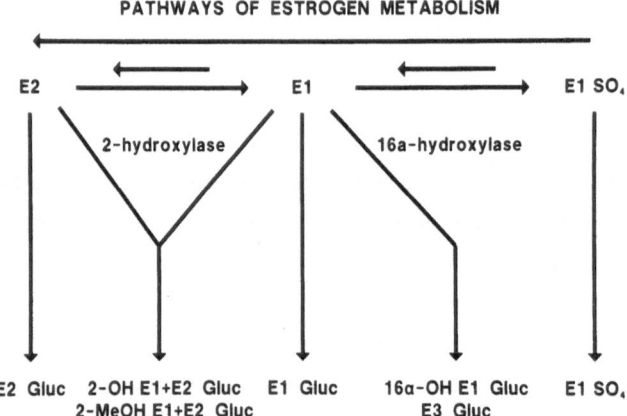

PATHWAYS OF ESTROGEN METABOLISM

Figure 1. Major pathways of metabolism of ^{3}H-estradiol. As shown, the glucuronidated matabolites of estradiol (E2) and estrone (E1) via the 2-hydroxylase pathway are combined to yield the products 2-OH E1 and 2-OH E2 and their metabolites, 2-methoxy E1(2-MeOH E1) and 2-methoxy E2 (2-MeOH E2). Similarly, as shown, the glucuronidated metabolites, 16a-hydroxy E1 (16a-OH E1) and estriol (E3) formed via the 16a-hydroxylase pathway are combined.

importance of the pathways[11]. This approach is labor intensive and requires multiple purification steps.

The second, or radiometric, approach involved the administration to the subject of estradiol specifically labeled with ^{3}H at either the 2, 16, or 17 position (Figure 2). Oxidation at a site so labelled will result in the loss of the ^{3}H into body water[12]. Measurement of radio-labeled body water would allow calculation of the percentage of estradiol undergoing oxidation at that specific site. This radiometric approach would not give data as to endogenous production, but is less labor intensive and does not require multiple purification steps.

The major pathways of estradiol metabolism measured by the first approach are noted in Figure 1. The metabolic clearance of estradiol and the ratios of estrone, estrone sulfate, and the glucuronides of estrone and estradiol were measured in the blood[11]. Because the concentrations of the radiolabeled products were below detection limits the ratios of both 2-hydroxy estrone and 2-hydroxy estradiol (the catechol estrogens) and the 16α-hydroxylated estrogens to estradiol were not measured. As shown, the products (2-

hydroxy estrone and 2-hydroxy estradiol) of the 2-hydroxylase pathways for estrone and estradiol are combined and also include the 2-methoxy metabolites of the 2-hydroxy estrogens. The data for the 16α-hydroxylated compounds, estriol and 16α-hydroxy estrone, are also combined, as shown. The sulfate conjugates, including estrone sulfate, were measured but were not important in relation to the glucuronides and will not be discussed further[11].

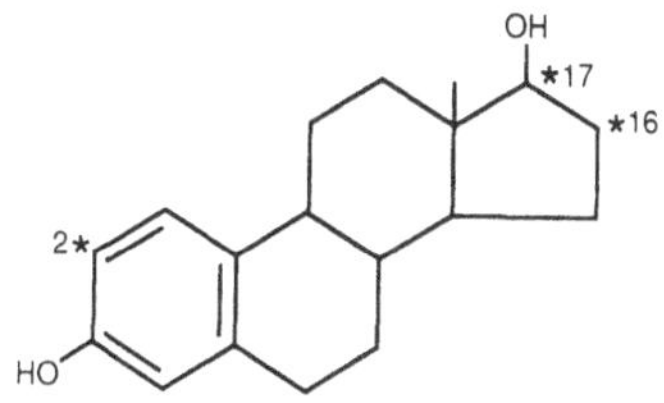

ESTRADIOL

Figure 2. The structure of estradiol showing the 3 sites which are labelled with ^{3}H in the radiometric approach.

Using the second, or radiometric, approach (Figure 2) the percentage of administered estradiol metabolized via oxidation at the C-17 position, the C-16 position, or the C-2 position was measured. At the C-16 position the metabolism involves the C-16α hydroxylase and at the C-2 position the 2-hydroxylase enzyme systems[12].

In normal women, the metabolism of estradiol and ethinyl estradiol characterized by measurements made from the blood pool is shown in Table 1. There is considerable metabolism of estradiol to estrone and subsequent sulfation of the estrone so that the major circulating metabolite after the administration of estradiol is estrone sulfate. The glucuronides of estrone and estradiol are present but, as noted above, the concentrations of the 2-hydroxylated and 16α-hydroxylated estrogens are too small to measure[11].

The metabolic clearance rate of estradiol is greater than that of ethinyl estradiol, a difference that cannot be explained by differences in the binding of these estrogens to sex hormone-binding globulin. The difference is probably due, in large part, to the lack of oxidation of ethinyl estradiol at the C-16 and C-17 positions and the relative lack of de-ethinylation[13-15]. The major circulating estrogen after the administration of ethinyl estradiol is the sulfate, but little attempt was made to identify circulating glucuronides.

The pattern of urinary metabolites is shown in Figure 3 and measurements in the urine do allow circulation of the percentage of administered estradiol that would be metabolized through the major pathways including glucuronidation, 2-hydroxylation and 16α-hydroxylation. For normal women in the follicular phase of the cycle, metabolism via 2-hydroxylation is quantitatively the most important pathway although considerable metabolism through 16α-hydroxylation also occurs[11].

Although all these data in blood and urine were obtained after the intra-venous administration of the radio-labelled estradiol, similar patterns were noted after the oral administration of radio-labeled estradiol[11]. However, after oral administration only about 10% of the radio-labelled estradiol was absorbed as estradiol into the blood and 90% was metabolized in the intestinal wall and liver before entering the systemic circulation. It should be noted that for orally administered ethinylestradiol most of the compound enters the blood as ethinylestradiol since there is a very small 'first pass' effect[15].

Table 1. Metabolism of estradiol and ethinyl estradiol measured in the blood pool while subjects were eating a standard Western diet.

	ESTRADIOL [a]	ETHINYL ESTRADIOL	
MCR[b] L/day	1,280		1,070 [c]
CR(E2,E1)[d]	0.31		-
CR(E2,E1SO4)	4.77	CR(EE2,EE2SO4)	6.48 [e]
CR(E2,E1Gluc)	0.23		-
CR(E2,E2Gluc)	0.27		-

[a] = Data from Longcope, et.al. (11).

[b] = MCR is metabolic clearance rate.

[c] = Data from Longcope and Williams (20).

[d] = CR(E2,E1) is the ratio of concentrations of estrone to estradiol after the injection of ^{3}H-estradiol.

[e] = Data from Bird and Clark (21).

Using the radiometric approach a quantitative estimation of the pathways is shown in Table 2. Oxidation at C-17 is the major route of metabolism, but this percentage would include that oxidized at C-2 and C-16[12]. The pattern of C-2 and C-16 oxidation is similar for both approaches although the actual quantities differ slightly.

The effects of dietary perturbation on the pathways of estrogen metabolism have been carried out using both tracer approaches with measurements in the blood, urinary glucuronide and body water pools. As shown in Table 3, the metabolic clearance rates and concentration (conversion) ratios were not altered when women were switched from a high fat, western style diet (40% of total calories as fat) to a low fat diet (25% of total calories

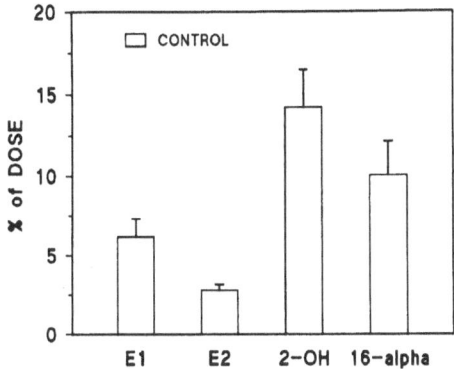

Figure 3. The percentages of the administered dose of ^{3}H-estradiol that are excreted in the urine as glucuronides. The 2-hydroxy and 3-methoxy compounds are shown as 2-OH and the 16a0hydroxy compounds are also combined. The subjects were normal young women who were eating a western-style, high fat diet.

as fat)[16]. In this study the concentration of estradiol in the blood and, hence the calculated blood production rates, were not altered. The fall in circulating estradiol in low fat-consumers, as reported in other studies[7,9], would, therefore, appear due to a decrease in the blood production rate of estradiol, since we detected no change in the metabolic clearance rate in our studies.

When measurements were made using urinary glucuronides, the low fat diet (Figure 4) resulted in a significant decrease in the 16α-hydroxylated estrogens and an increase, although not significant, in the 2-hydroxylated estrogens. Musey, et. al.[17] reported (Table 4) that when chimpanzees were placed on a high fat diet there was a significant increase in 16α-hydroxylase activity and a decrease in the 2-hydroxylase activity. Thus, it would appear that the 2-hydroxylase activity is stimulated by a low fat diet and inhibited by a high fat diet, with reciprocal changes in the 16α-hydroxylase activity. All the above data were

Table 2. Pathways of estradiol matabolism in women as measured by the radiometric approach.

^{3}H Site	Percent Oxidized
C-2	37.3%[a]
C-16	13.7%
C-17	85.4%

[a]=Data from Fishman, et.al (12).

obtained following IV administration of the radio-labelled estradiol, however, in our study the low fat diet caused similar changes in the metabolism of orally administered estradiol[16]. The fractional absorption of the orally administered estradiol was not altered by the change in diet. The failure to find an effect of diet on blood pool measurements, however, is not incompatible with the urinary glucuronide findings.

Another dietary perturbation that affects estradiol metabolism was reported by Anderson et. al. (Table 5)[18]. These workers found that high protein, low carbohydrate diets resulted in greater oxidation of estradiol at the C-2 position than did a low protein, high carbohydrate diet[18]. They ascribed this to a stimulation of the specific cytochrome P-450 mixed function oxidase that is involved in 2-hydroxylation. Interestingly, there was no effect of the different diets on oxidation at the C-16 position.

Table 3. The effects of a low fat diet on the metabolism of ^{3}H-estradiol as measured in the blood pool.

	HIGH FAT DIET [a]	LOW FAT DIET
MCR L/day	1,280	1,290
CR(E2,E1)	0.31	0.36
CR(E2,E1SO4)	4.77	4.79
CR(E2,E1Gluc)	0.23	0.31
CR(E2,E2Gluc)	0.27	0.16

[a] = Data from Longcope, et.al. (16).

Subsequently, Michnovicz and Bradlow[19] made the very interesting observation that the administration of an indole, specifically indole-3-carbinol, was associated with increased oxidation of estradiol at the C-2 position. The increase of 2-hydroxylation was ascribed to the induction of the specific cytochrome P-450-dependent mixed function oxidase by the indole, an effect similar to that of the high-protein, low-carbohydrate diet (Table 6).

The effect of a high fiber diet on estradiol metabolism has recently been studied in association with Drs. Gorbach, Goldin, Dwyer, and Woods (unpublished data). When women were placed on diets high in fiber content, the metabolism of estrogen as measured in the blood pool was not altered. However, we did find alterations in urinary glucuronides (data not shown) that resembled those seen with a low fat diet but to date these changes with a high fiber diet have not been significant.

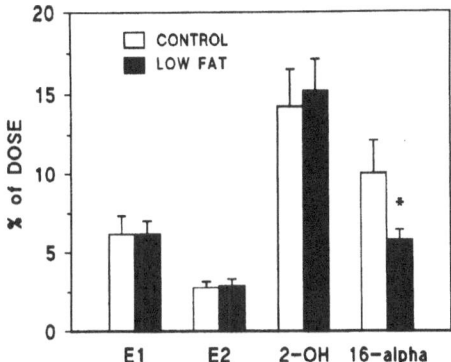

Figure 4. The effects of a low-fat diet on the percentage of ^{3}H-estradiol excreted in the urine as specific matabolites. Comparison is made with the percentage excreted when the same normal young women were on a high fat diet. *=p<0.01 low fat compared with high fat diet.

Thus, estradiol metabolism can be altered by diets low in fat or high in protein and low in carbohydrate. The administration of an indole, present in cruciferous vegetables also alters estrogen metabolism and presumably these changes are due, in large part, to induction of the cytochrome P-450 mixed function oxidase.

Whether such dietary perturbations alter the metabolism of ethinyl estradiol is conjectural, at best. There are few, if any, data available, but since the 2-hydroxylase pathway is involved in ethinyl estradiol metabolism, it is probable that this pathway would be stimulated or inhibited by diet in the same manner as for estradiol. However, it is doubtful whether dietary perturbations would influence the absorption of orally administered ethinyl estradiol, and hence the circulating levels and clinical effectiveness probably would be unchanged. This remains unproven, however, and merits further study.

Table 4. The effects of a high fat diet on the metabolism of estradiol in chimpanzees as measured by the radiometric technique[a].

	Percent oxidation	
^{3}H-site	Control	High Fat
C-2	31.6%	20%
C-16	3.2%	19.6%

[a] = Data from Musey et.al.(17).

Table 5. The effects of a diet high in protein or high in carbohydrate on the metabolism of estradiol as measured by the radiometric technique[a].

^{3}H-site	Percent oxidation	
	High Protein	High CHO
C-2	44%	33%
C-16	⁻13.5%	⁻12.5%

[a] = Data from Anderson, et.al. (18).

Table 6. Effect of indole-3-carbinol ingestion on the metabolism of estradiol as measured by the radiometric approach.

Site of ^{3}H	Percent Oxidized	
	Control	Indole-3-carbinol
C-2	29.3%[a]	45.6%[a]
C-16	13.7%[b]	nr [c]
C-17	85.4%	nr

[a] = Data from Michnovicz and Bradlow (19)

[b] = Data from Fishman, et.al. (12)

[c] nr = not reported

References

1. Armstrong B, Doll R. 1975. Environmental factors and cancer incidence and mortality in different countries, with special reference to dietary practices. Int J Cancer 15:617.
2. Miller AB, Gori GB, Kunze M, Grahm S, Reddy BS, Hirayama T, Weisburger J. 1980. Nutrition and cancer. Prev Med 9:189.
3. Miller AB, Kelly A, Choi NW, Matthews V, Morgan RW, Munan L, Burch JD, Feather J, Howe GR, Jain M. 1978. A Study of diet and breast cancer. Am J Epidemiol 107:499.
4. Kelsey JL. 1979. A review of the epidemiology of human breast cancer. Epidemiol Rev 1:74.
5. Goldin BR, Adlercreutz H, Gorbach SL, Warram JH, Dwyer JT, Swenson L, Woods MN. 1982. Estrogen excretion patterns and plasma levels in vegetarian and omnivorous women. N Engl J Med 307:1542.
6. Gray GE, Williams P, Gerkins V, Brown JB, Armstrong B, Phillips R, Casagrande JT, Pike MC,Henderson BE. 1982. Diet and hormone levels in Seventh-Day Adventist teenage girls. Prev Med 11:103.
7. Rose DP, Boyar AP, Cohen C, Strong LE. 1987. Effect of a low-fat diet on hormone levels in women with cystic breast disease. I. Serum steroids and gonadotropins. J Natl Cancer Inst 78:623.
8. Hagerty MA, Howie BJ, Tan S, Schultz TD. 1988. Effect of low- and high-fat intakes on the hormonal milieu of premenopausal women. Am J Clin Nutr 47:653.
9. Prentice R, Thompson D, Clifford C, Gorbach S, Goldin B, Byar D. 1990. Dietary fat reduction and plasma estradiol concentration in healthy postmenopausal women. J Natl Cancer Inst 82:129.
10. Baird D, Horton R, Longcope C, Tait JF. 1968. Steroid prehormones. Perspect Biol Med 11:384.
11. Longcope C, Gorbach S, Goldin BM, Woods M, Dwyers J, Warman J. 1985. The metabolism of estradiol: oral compared to intravenous administration. J Steroid Biochem 23:1065.
12. Fishman J, Bradlow HL, Schneider J, Anderson KE, Kappas A. 1980. Radiometric analysis of biological oxidations in man: Sex differences in estradiol metabolism. Proc Natl Acad Sci U S A 77:4957.
13. Cargill DI, Steinetz BG, Gosnell E, Beach VL, Meli A, Fujimoto GI, Reynolds BM. 1969. Fate of ingested radiolabeled ethynylestradiol and its 3-cyclopentyl ether in patients with bile fistulas. J Clin Endocrinol Metab 29:1051.
14. Abdel-Aziz MT, Williams KIH. 1970. Metabolism of radioactive 17a-ethinylestradiol by women. Steroids 15:695.
15. Longcope C, Williams KIH. 1977. Ethynylestradiol and mestranol: their pharmacodynamics andeffects on natural estrogens. In: SGarattini - #HWBerendes (eds) Pharmacology of Steroid Contraceptive Drugs. Raven Press, New York, p 89.
16. Longcope C, Gorbach S, Goldin B, Woods M, Dwyer J, Morrill A, Warram J. 1987. The effect of a low fat diet on estrogen metabolism. J Clin Endocrinol Metab 64:1246.
17. Musey PI, Collins DC, Bradlow HL, Gould KG, Preedy JRK. 1987. Effect of diet on oxidation of 17b-estradiol in vivo. J Clin Endocrinol Metab 65:792.
18. Anderson KE, Kappas A, Conney AH, Bradlow HL, Fishman J. 1984. The influence of dietary protein and carbohydrates on the principal oxidative biotransformations of estradiol in normal subjects. J Clin Endocrinol Metab 59:103.
19. Michnovicz JJ, Bradlow HL. 1990. Induction of estradiol metabolism by dietary indole-3-carbinol in humans. J Natl Cancer Inst 82:947.
20. Longcope C, Williams KIH. 1975. The metabolism of synthetic estrogens in non-users and users of oral contraceptives. Steroids 25:121.
21. Bird CE, Clark AF. 1973. Metabolic clearance rates and metabolism of mestranol and ethinylestradiol in normal young women. J Clin Endocrinol Metab 36:296.

DIET, CIRCULATING ESTROGEN LEVELS AND ESTROGEN EXCRETION

Barry R. Goldin[1]

There have been a number of reports of increased frequency of breakthrough bleeding and pregnancies in women taking oral contraceptives and antibiotics simultaneously (1-4). One affect of the possible mechanism for this is the ability of antibiotics to interfere with the enterohepatic circulation of estrogens (5). Diet may also influence estrogen cycling and therefore impact on the effectiveness of oral contraceptives. The relationship between diet and plasma estrogen levels and excretion patterns are the subject of this report.

In women of reproductive age, the primary source of circulating estrogens is the ovary, although estrone production arises in part from the extraglandular aromatization of androstenedione (6). In postmenopausal women, significant amounts of estrogens are synthesized in extraglandular sites (7). Following synthesis, the estrogens circulate in the blood in the unconjugated form and are transported to the target tissues and the liver. The major biologically active estrogens are the unconjugated (free) estrogens, of which estradiol is the most active (8). Estrone is present in nearly equal amounts as estradiol. Estrone possesses weak activity of its own but can be converted intracellularly to estradiol. Estradiol is present in only small quantities in the free form and has weak estrogenic activity (9).

In the liver, and also in other tissues, the estrogens are metabolized. The major pathway of metabolism is from estradiol to estrone (10). Estrone is then metabolized further to estriol or to the catechol estrogens which are then conjugated, primarily as glucuronides and sulfates (11). Greater than 50% of the metabolism and conjugation of

B. GOLDIN • Tufts University School of Medicine, Department of Community Health, Boston, MA

Steroid Contraceptives and Women's Response,
Edited by R. Snow and P. Hall, Plenum Press, New York, 1994

153

estrogens occurs in the liver. Some of these conjugates reenter the bloodstream, where they are transported to the kidney and excreted. The glucuronide conjugates are excreted in the urine more rapidly than are the sulfates. The estrogen sulfates are excreted only slowly and can be hydrolyzed in tissues and act as a source of biologically active estrogen (12).

Approximately 50% of the estrogen conjugates, which enter or are formed in the liver, are excreted in the bile, pass into the intestine, and are hydrolyzed by bacteria (13). They are then either reabsorbed or excreted in the feces as the free steroid. The reabsorbed estrogens enter into the portal system; they are conjugated by the splanchnic tissues and either pass into systemic circulation or pass into bile. The excretion of estrogens into bile with subsequent hydrolysis and reentry into the bloodstream comprises the enterohepatic circulation. The hydrolysis of the estrogen-glucuronides is accomplished by the intestinal bacterial enzyme beta-glucuronidase. The intestinal microflora also perform reductive and oxidative reactions on estrogens and androgens which could alter the spectrum of metabolites reabsorbed from the intestine (14).

Experiments from the laboratory of Adlercreutz and colleagues first showed that bacterial activity in the bowel can be important in determining estrogen status in women. Administration of oral ampicillin to pregnant women resulted in a 34% decrease in urinary estrogen excretion and a 6-fold increase in the excretion of fecal estrogen (15,16). The concentration of conjugated forms of estrogen in the feces increased 60-fold (15).

A positive correlation between body weight and plasma levels of estrone and estradiol has been reported by several workers (17-20). The level of urinary estrogen excretion is low in undernourished women (21,22), whereas obese women have higher plasma levels of estrogen, presumably because estrogens are converted from androgens in their fat tissue (23,24).

HIGH FAT INTAKE
(INCREASES INTESTINAL ABSORPTION VIA
INCREASED BACTERIAL DECONJUGATION)

$\downarrow$

INCREASED ENTEROHEPATIC CIRCULATION
OF STEROID HORMONES

$\downarrow$

INCREASED PLASMA AND URINE ESTROGEN LEVELS
AND DECREASED FECAL ESTROGEN LEVELS

Figure 1. Hypothesis 1: relationship between high dietary fat intake and estrogen status.

HIGH FIBER INTAKE
(DECREASES INTESTINAL ABSORPTION OF ESTROGENS
VIA SEQUESTRATION AND/OR LOWER BACTERIAL
DECONJUGATING ENZYME ACTIVITY)

DECREASED ENTEROHEPATIC CIRCULATION
OF STEROID HORMONES

LOWER PLASMA AND URINARY ESTROGENS AND
INCREASED FECAL EXCRETION

Figure 2. Hypothesis 2: relationship between high dietary fiber intake and estrogen status.

Other studies have shown that the composition of the diet also influences circulating estrogen concentration and excretion of estrogen hormones in women. Armstrong et al. (25) studying U.S. vegetarian and omnivore women, found differences in serum prolactin and sex hormone binding globulin as well as urinary estrogens. Vegetarians have lower urinary levels of estriol and total estrogens, lower plasma prolactin and higher plasma sex hormone binding globulins. In another study of plasma estrogen levels in vegetarian and nonvegetarian women, 14 premenopausal seventh-day Adventist women were compared with 9 premenopausal omnivores (26); the vegetarian women consumed significantly less fat, especially saturated fat, than the omnivores. Plasma levels of estrone and estradiol were found to be lower in the vegetarians. Rose et al. (27) examined the effect of a low fat diet on serum estrogens in 16 premenopausal patients with cystic breast disease and cyclic mastalgia. Serum samples were taken prior to the dietary intervention at which time the women ate an average of 69 grams of fat (35% of total calories). Serum was then collected 2 and 3 months thereafter at which time the average fat consumption was 32 grams (21% of calories). After 3 months on this low-fat diet there were significant reductions in luteal-phase serum total estrogens, estradiol and estrone. Prentice et al. (28) studied 73 healthy postmenopausal women before and after 10-22 weeks of participation in a low-fat diet intervention. These women reduced dietary fat from 40% of calories to 20%. There were significant reductions in plasma total estradiol and free (bioavailable) estradiol at the end of the intervention.

In our laboratory we have conducted a series of studies to determine whether diet has an effect on estrogen excretion and metabolism. These studies have concentrated either on populations that have different dietary intakes of fat and fiber, volunteers fed prepared diets containing different amounts and types of fat or fiber, and an intervention study in which women over 45 years of age were counselled on lowering their fat intake. The latter study was part of the "Women's Health Trial," (28).

RESULTS

Two of the hypotheses that our studies are based on are presented in figures 1 and 2. Figure 1 presents a rationale for some of the observed effects of a high fat diet on plasma estrogen levels and estrogen excretion. There have been a number of studies in humans and animals which have shown that a low fat intake results in a significant reduction of fecal bacterial beta-glucuronidase activity (12,29,30). Therefore it may follow that individuals eating high fat diets would have increased estrogen deconjugation and a resultant increase in intestinal absorption of estrogens. This would result in higher plasma estrogen concentrations and lower amounts of estrogen excreted in the feces. In contrast, a high fiber diet (see figure 2) either traps estrogens in the intestine and/or by volume dilution reduces beta-glucuronidase concentration, in either manner decreasing re-absorption of estrogens from the intestine. This would result in higher fecal excretion and lower concentrations of estrogens in the plasma.

The initial study performed in our laboratory involved 10 vegetarian and 10 nonvegetarian premenopausal women (12). Table 1 presents the demographic and dietary data for these two groups. Blood, urine and feces were collected on four occasions, at intervals of approximately four months, during the mid-follicular phase of the menstrual cycle. In table 2 the data for the excretion of fecal estrogens are presented for the two groups of women. The vegetarians excreted significantly higher amounts of all three estrogens measured in the feces. In table 3 the amount of several important estrogens excreted in urine over a 24 hour period is shown for vegetarians and omnivores. The omnivores excreted significantly higher estriol and 37% higher estriol-3-glucuronide. Estriol undergoes an extensive enterohepatic circulation and is a good marker for this process; similarly estriol-3-glucuronide is formed in the intestinal tissue after absorption of estriol from the lumen and is therefore also a good marker for intestinal absorption (31). In table 4 the plasma estrone and estradiol concentrations are shown for vegetarians and omnivores. The omnivores had 17% higher estrone and 23% higher estradiol, but the differences did not achieve statistical significance. Several other studies (28,32) have found significant diet related differences in plasma estrogens, with a magnitude of the change similarly around 20%, however, these studies had larger subject populations. In figure 3 the negative correlation between plasma estrogen concentration and fecal estrogen excretion for the study population is shown. This correlation was significant at p<0.003. These data demonstrate that a decreased enterohepatic as reflected in higher fecal estrogen excretion results in lower amounts of circulating estrogen.

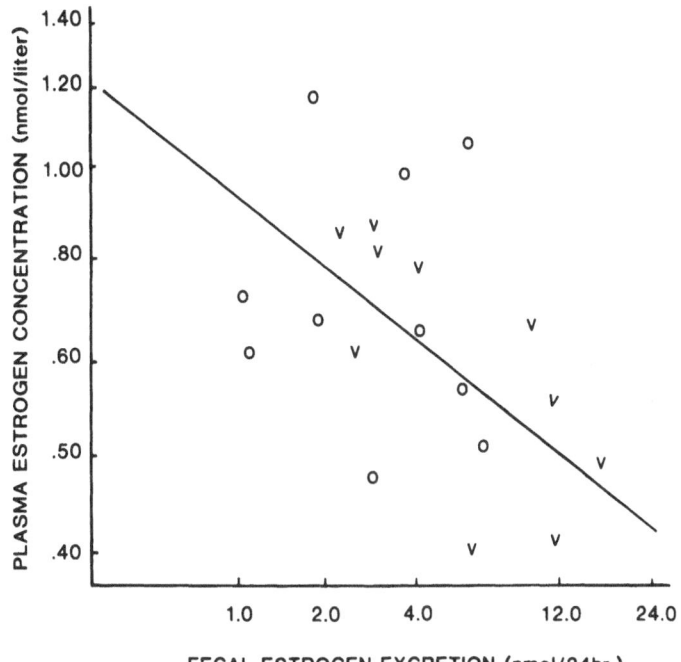

Figure 3. Log-Plot of the Correlation between 24-hour Fecal Excretion of Estrogen (sum of estrone, estradiol, and estriol) and the plasma concentration of estrogen (sum of estrone and estradiol). Vegetarians (V) and omnivores (O).

Upon completion of the study described above a new study was initiated in which pre-and postmenopausal Caucasian omnivores, Lacto-ovo vegetarians, vegans and recent Asian immigrants to Hawaii (primarily Vietnamese) participated. In the previous study the vegetarians were primarily lacto-ovo vegetarians, which indicated they consume diary products and fish. Vegans, in contrast, do not eat any animal products. The vegans also ate a higher fiber diet with an average intake of approximately 30 grams per day. In Table 5 the dietary consumption data for the pre and postmenopausal omnivores and Asians who participated in this study are presented.

In table 6 the serum values are shown for the pre and postmenopausal omnivores, lacto-ovo vegetarians and the vegans. Among premenopausal women, vegetarians had 12.5% lower serum estrone compared to omnivores; this difference was not statistically significant. However, the vegans had 22% lower estrone when compared to omnivores. For the postmenopausal women serum estrone and estradiol concentration was 30 and 28% lower. When compared to omnivores these changes were not significant due to the small sample size, however, the magnitude of the change was large, suggesting that postmenopausal vegans have lower circulating estrogen. In Table 7 fecal estrogen excretion values are shown for premenopausal Caucasian-omnivores and East Asian women and in Table

Table 1. Demographic and dietary data on 10 omnivorous and 10 vegetarian women.

	Omnivores		Vegetarians	
DEMOGRAPHIC DATA (arithmetic mean±S.E.)				
Age (yr)	25.5±1.0		26.5±0.9	
Height (cm)	166.8±2.7		163.7±2.4	
Weight (kg)	63.8±2.6		59.9±2.9	
Body-mass index: weight (kg)/height(cm)^2	22.7±0.5		22.6±1.0	
DIETARY DATA (daily consumption)[*]				
Kilocalories	1572	(1496-1652)	1649	(1625-1734)
Protein (g)	62	(59-65)	55	(52-58)
Animal Protein (%)	43[1]		7	
Dietary Fiber (g)	12[1]	(11-13)	28	(25-33)
Dietary Fat				
Total fat	70	(63-74)	56	(53-60)
Total fat (g/1000 cal)	44[2]	(42-46)	34	(31-37)
Saturated fat (g)	27[3]	(25-29)	17	(16-19)
Saturated fat (g/1000 cal)	17[2]	(16-18)	10	(9-12)
Calories as fat (%)	40		30	

[*]With the exception of total fat and calories from fat, all dietary data are expressed as geometric means, with the range of S.E. in parentheses.
[1]$p<0.001$.   [2]$p<0.05$.   [3]$p=0.07$.

Table 2. Fecal excretion of estrogens in 10 omnivorous and 10 vegetarian women[*]

Hormone	Omnivores		Vegetarians	
	nmol/24 hours			
Estrone	0.83	(0.70-0.99)	1.96	(1.68-2.29)[2]
Estradiol	0.61	(0.52-0.72)	1.52	(1.30-1.78)[1]
Estriol	0.72	(0.63-0.83)	1.72	(1.50-1.98)[2]
Total Estrogen (estrone, estradiol, estriol)	2.33	(2.01-2.70)	5.40	(4.70-6.21)[1]

[*]Values are presented as geometric means with S.E. ranges in parentheses. Calculations are based on four collections per subject. [1]$p<0.03$ [2]$p<0.01$

Table 3. Urinary Excretion of estrogens in 10 omnivorous and 10 vegetarian women.[*]

Hormone	Omnivores		Vegetarians	
	nmol/ 24 hours			
Estrone	15.3	(13.8-17.0)	16.8	(15.0-18.8)
Estradiol	9.3	(8.6-10.0)	9.3	(8.5-10.2)
Estriol	21.0	(19.1-23.1)	13.2	(11.6-15.0)[1]
Estriol-3-glucuronide	38.5	(36.0-41.2)	28.0	(24.8-31.6)

[*]Values are presented as geometric means with S.E. ranges in parentheses. Calculations are based on four collections per subject. [1]$p<0.05$.

Table 4. Plasma estrogen and androgen levels in 10 omnivorous and 10 vegetarian
women[*].

Hormone	Omnivores	Vegetarians
	nmol/liter	
Estrone	0.40 (0.36-0.44)	0.34 (0.32-0.36)
Estradiol	0.32 (0.29-0.36)	0.26 (0.23-0.29)

[*]Values are presented as geometric means with S.E. ranges in parentheses. Calculations are based on four
collections per subject.

Table 5. Daily mean nutrient intake for Caucasian Americans and Oriental immigrant
women

	Daily Nutrient Intake[*]			
	Premenopausal		Postmenopausal	
Nutrient Oriental	Caucasian (omnivore) n=10	Oriental n=12	Caucasian (omnivore) n=11	n=11
Kilocalories	1982 (1875-2094)	1534 (1404-1675)	1746 (1696-1769)[1]	1286(1188-1391)
Diet. Fiber(g)	11.0 (9.7-12.4)	15.2 (13.5-17.0)	15.7 (14.0-17.6)	19.5 (16.1-23.6)
Total fat(g)	89.1 (83.3-95.3)[1]	37.3 (33.1-43.0)	72.4 (68.7-76.2)[1]	27.5 (24.0-31.4)
saturated fat(g) calories	36.4 (33.7-39.3)[1]	12.9 (11.4-14.6)	26.0 (24.5-27.5)[1]	9.0 (7.9-10.1)
as fat(%)	40.4 (39.1-41.7)[1]	21.8 (20.6-23.0)	38.2 (36.5-39.9)[1]	19.2(17.9-205)
Protein (g)	72.0 (67.0-77.3)	76.9 (70.1-84.3)	73.8 (70.5-77.2)[1]	59.4 (54.2-65.1)
Carbohydrate(g)	198.8 (182.1-216.9)	216.2 (197.4-236.9)	185.7 (172.4-200.0)	200.9(188.4-214)
fat/fiber(g/g)	7.7 (6.7-8.7)[1]	2.4 (2.1-2.7)	4.6 (4.0-5.2)[1]	1.4 (1.1-1.7)

[*]All dietary data is expressed as geometric mean and standard error range.   [1]p<0.05, compared to value from
corresponding age group.

Table 6. Serum estrogen concentrations in omnivores, lacto-ovo vegetarians, and vegetarians

Group	Number of determinations	Estrone nmol/liter	Estradiol
Premenopausal			
Omnivore	20	.322±.096[a]	.251±.060[a]
Lacto-ovo vegetarian	12	.282±.086	.255±.088
Vegan	15	.251±.087[a]	.203±.062[b]
Postmenopausal			
Omnivore	16	.162±.062	.115±.040
Lacto-ovo vegetarian	4	.155±.015	.111±.017
Vegan	5	.113±.045	.071±.052
			±.052

Mean   ± standard error   a:b p<0.05

Table 7. Fecal excretion of estrogens in Caucasian American and Oriental immigrant women.

Hormones[*]	Premenopausal		
	Caucasians (n=10)	Orientals (n=12)	p
Estrone	1.10 (o.82-1.46)	2.06 (1.68-2.51)	<0.05
Estradiol	0.54 (0.40-0.69)	1.68 (1.36 -2.06)	<0.02
Estriol	0.93 (0.66-1.30)	3.50 (2.95-4.13)	<0.02
Total Estrogen	2.64 (1.95-3.56)	7.53 (6.38-8.88)	<0.02

[*]Values are presented as geometric means of nmol/24hour with SE ranges in parentheses.

Table 8. Plasma estrogen levels in pre- and postmenopausal Caucasian American and Oriental immigrant women

Hormone[*]	Premenopausal		Postmenopausal	
	Caucasian (n=10)	Oriental (n=12)	Caucasian (n=10)	Oriental (n=9)
Estrone	0.32 (0.30-0.34)	0.24 (0.22-0.25)[1]	0.14 (0.13-0.15)	0.13 (0.10-0.15)
Estradiol	0.25 (0.24-).26)	0.14 (0.12-0.15)[2]	0.10 (0.09-0.11)	0.03 (0.02-0.04)[2]
Estrone plus Estradiol	0.57 (0.54-0.60)	0.39 (0.36-0.42)[2]	0.24 (0.22-0.26)	0.17 (0.14-0.21)[3]

[*]Values are presented as geometric means of nmol/L with SE ranges in parentheses.
[1]p<0.02   [2]p<0.001   [3]p=0.05

8 the plasma concentrations for pre and postmenopausal East Asian and Caucasian women are shown. The East Asian women excreted significantly higher estrogen in the feces and this was seen equally for estrone, estradiol and estriol (Table 7).

The premenopausal Asian Women had significantly lower plasma estrone, estradiol, and estrone plus estradiol and the postmenopausal Oriental women had significantly lower estradiol and estrone plus estradiol. These differences may in part result from the drastically different diet these women consume.

In figure 4 the positive correlation between fat consumption and serum estradiol concentration for the omnivore, lacto-ovo and vegan is shown. These data suggest that fat does influence plasma estrogen concentrations in populations eating different fat levels for long periods of time. These data do not exclude effect by dietary fiber on circulating estrogen levels. In this regard data will be presented below which indicates fiber does have an effect.

In Figures 5,6,and 7, the relationship between total fat, saturated fat and dietary fiber and plasma estrogen (estriol or estrone) are shown for 10 omnivores and 12 Asian premenopausal women. There was a significant positive correlation between total fat and plasma estradiol (p<0.001) and saturated fat and plasma estrone (p=0.012). In addition, a significant negative correlation was found between dietary fiber and plasma estradiol. These findings further confirm the relationship between diet and circulatory estrogen concentration.

In recent years our laboratory has conducted studies in which premenopausal volunteers have been fed defined diets prepared and eaten in a research metabolic unit in

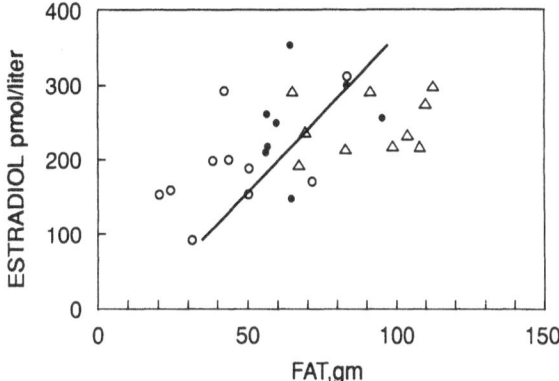

Figure 4. Correlation between fat intake and serum estradiol. All women were premenopausal. △ omnivores, ○ lacto-ovo vegetarians, ●vegans; p=0.005, R=0.511.

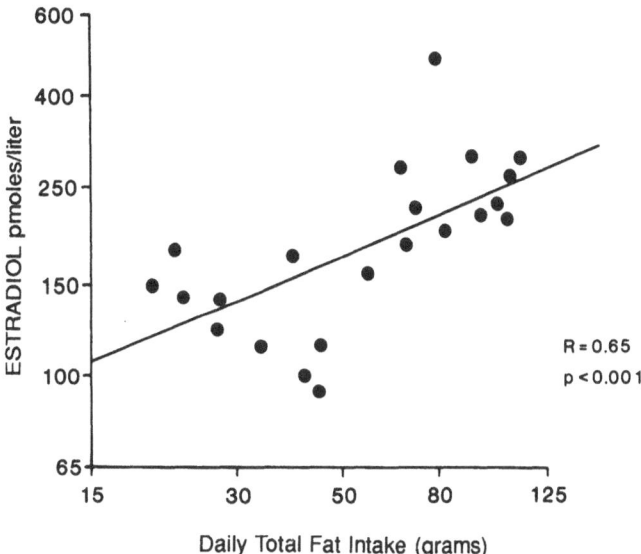

Daily Total Fat Intake (grams)

Figure 5. The relationship between total daily fat intake from the diet and serum estradiol levels. The 22 points represent mean values for fat intake and estradiol for four collections obtained from 10 Caucasians and two collections from the 12 Oriental immigrants. The graph is a log-log plot and the numerical values are the antilogs.

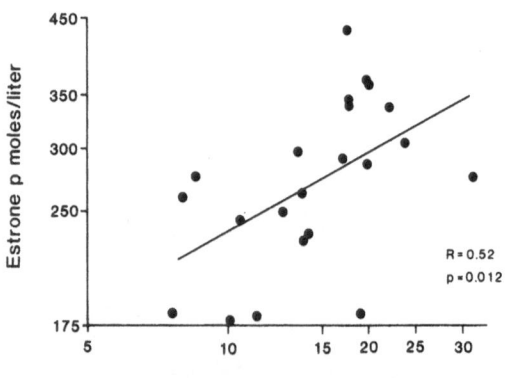

Figure 6. The relationship between daily saturated-fatty acid intake from the diet and serum estrone levels. For further details see figure 5.

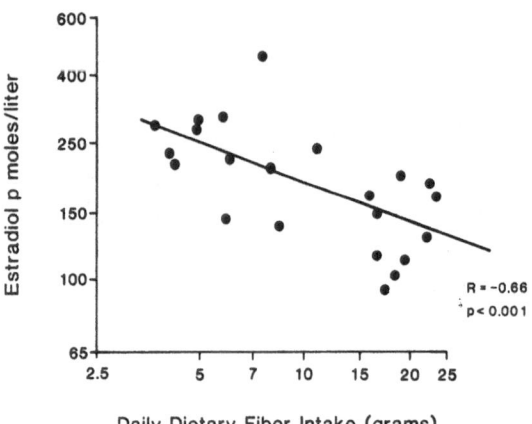

Figure 7. The relationship between daily fiber intake and serum estradiol. For further details see figure 5.

Table 9. Diet and estrogen study summary of dietary protocols.

	Control	Protocol 1	2	3	4	5	6	7
Premenopausal (n)		7	9	10	9	6	6	5
Fat (%)	40	25	25	25	20	20	40	20
P:S Ratio	0.5	0.5	0.5	1.0	1.0	1.0	0.5	1.0
P:S:M	1:2:2	1:2:2	1:2:2	1:1:1	1:1:1	1:1:1	1:2:2	1:1:1
Cholesterol (mg)	400	400	400	200	150	400	400	150
Fiber (g)	12	12	40	40	40	40	40	12

Consumption of Control and Experimental Diets

Table 10. The effect of Fiber Repeated Measures Analysis of Co-Variance

DEPENDENT VARIABLE	CORRELATION	P VALUE
E^1	negative	.15
E_1-SO^4	negative	.009
E_2	negative	.12
SHBG	negative	.09

FOLLICULAR PHASE

Table 11. The effect of Fat Repeated Measures Analysis of Co-Variance

	FOLLICULAR PHASE	
DEPENDENT VARIABLE	CORRELATION	P VALUE
E_2	negative	.03
Free E_2	negative	.03
Testosterone	negative	.12

Table 12. The effect of P:S Ratio Repeated Measures Analysis of Co-Variance

	FOLLICULAR PHASE	
DEPENDENT VARIABLE	CORRELATION	P VALUE
Androstendione	positive	.02
SHBG	positive	.04
Prolactin	positive	.006
Cycle Length	positive	.008

a facility devoted to human nutritional studies. In Table 9 the dietary protocols in terms of fat, fiber, polyunsaturated to saturated fat ratio (P/S ratio) and dietary fiber are presented for the baseline and experimental diets. The diets were designed to determine which of these dietary components alter circulating estrogen levels. In a previous paper (33) we reported that 17 premenopausal women following either protocol 4 or 5 had a significant reduction in serum estrone sulfate when fed the experimental diet for two months compared to levels measured after one month on the control diet. The concentration declined in the serum from 2.11 n mol/L to 1.29 n mol/L. The decline was highly significant with a P value of less than 0.001. The estrone sulfate serum concentration for all the premenopausal women on the control diet was 489.80 pg/ml (440.5-544.7) (geometric mean and standard error range). The same women eating the various experimental diets had a concentration

Table 12. The effect of P:S Ratio Repeated Measures Analysis of Co-Variance

	FOLLICULAR PHASE	
DEPENDENT VARIABLE	CORRELATION	P VALUE
E_2	negative	.01
Free E_2	negative	.11
SHBG	negative	.15
Progesterone	negative	.09

of 343.77 pg/ml (315.51-374.87). This difference is also statistically significant. In table 10,11,12 and 13 the results from an analysis of co-variance is shown for various outcome measurements for all subjects analyzed to date on the seven dietary protocols. Only significant changes or trends toward significance are presented. Increased fiber intake was significantly correlated with the decreased estrone (E_1) - sulfate noted above. Increased fiber intake appears to reduce serum estrone, estradiol and sex hormone binding globulin (SHBG). Increased dietary fat intake did not appear to have any effect on circulating estrogen levels, however, it did increase androstenedione, SHBG and prolactin concentrations in serum. An increase P:S ratio was related to decreased estradiol and free estradiol concentration in the plasma and increased dietary cholesterol also resulted in decreased

estradiol levels. These data indicate there is a complex relationship between estrogen and diet, however, it appears that increased fiber in the diet and elevated polyunsaturated fat are responsible for lower estrogen and total fat has little effect in volunteers eating a defined diet for a relatively short duration (two months).

CONCLUSIONS

Studies conducted in our laboratory have shown that populations eating different diets have different plasma estrogen concentrations which is accompanied by altered estrogen excretion patterns. Individuals eating high fat - low fiber diets have higher circulating estrogen and a higher urinary to fecal excretion ratio when compared to individuals eating high fiber - low fat diets. Studies in which defined diets varying in fat, fiber, P/S ratio and cholesterol are eaten for up to two months indicate that high intake of fiber and an elevated P/S ratio lower circulating estrogen concentration. These findings may also impact on the effective dose of oral contraceptives delivered to the target tissue. These data suggest that populations eating low fat-high fiber diets may have an increased fecal excretion of the active estrogen component of the contraceptive and a lower level of tissue response.

References

1.   Dossetor J. Letle: Drug interactions with oral contraceptives. *Br Med J* 4:467-468. 1975.
2.   Mumford JP. Letle: Drugs affecting oral contraceptives. *Br and J* 2:333-334. 1974.
3.   Stockley I. Interactions with oral contraceptives. *Pharm J* 216:140-143. 1976.
4.   Bolt HM, Bolt M, Kappus H. Interaction of rifampicin treatment with pharmacokinetics and metabolism ethynlestradiol in man. *Acta Endocrinol* 85:189-190. 1977.
5.   Seevson L, Goldin B, Gorbach SL. Effect of antibodies on fecal/urinary excretion of ethinyl estradiol, on oral contraceptives. *Gastroenterology* 78(pt2):1332. 1980.
6.   Baird DT, *et. al.* Steroid dynamics under steady state conditions. *Recent Prog Horm Res* 25:611-664. 1969.
7.   Longcope C, Kato T, Horton R. Conversion of blood androgens to estrogens in normal adult men and women. *J Clin Invest* 48:2191-2201. 1969.
8.   Yen SSC. The human menstrual cycle. In: *Reproductive Endocrinology.* SSC Yen, RB Jaffe, (eds.). pp. 126-151. Saunders: Philadelphia. 1978.
9.   Rotti K, *et. al.* Estriol concentrations in plasma of normal, non-pregnant women. *Steroids* 25:807-816. 1975.
10.   Fishman J, Bradlow HL, Gallagher TF. Oxidative metabolism of estradiol. *J Biol Chem* 235: 3104-3107. 1960.
11.   Eriksson H, Gustafsson JA. Excretion of steroid hormones in adults. *Eur J Biochem* 18:146-150. 1971.
12.   Goldin BR, *et. al.* Estrogen excretion patterns and plasma levels in vegetarian and omnivorous women. *New Eng J Med* 207:1542-1547. 1982.
13.   Sandberg AA, Slaunwhite WR, Jr. Studies on phenolic steroids in human subjects. 2. The metabolic fats and hepatobiliary-enteric circulation of ^{14}C-estrone and ^{14}C-estrone and ^4C-estradiol in women. *J Clin Invest* 36:1266-1278. 1957.

14. Lombardi P, *et. al.* Metabolism of androgens and estrogens by human fecal microorganisms. *J Steroid Biochem* 9:795-801. 1978.

15. Adlecreutz H, *et. al.* Intestinal metabolism of estrogens. *J Clin Endocrinol Metab* 43:497-505. 1976.

16. Adlecreutz H, *et. al.* Effect of ampicillin administration on the excretion of twelve estrogens in pregnancy urine. *Acta Endocrinol* 80:551-557. 1975.

17. Bates GW, Whitworth NS. Effects of obesity on sex steroid metabolism. *J Chron Dis* 35:893-896. 1982.

18. Krischner MA, *et. al.* Obesity, androgens, estrogens and cancer risk. *Cancer Res* (Suppl.) 42: 3281S-3285S. 1982.

19. Kirschner MA, Ertel N, Schneider G. Obesity, hormones and cancer. *Cancer Res* 41:3711-3717. 1981.

20. Zumoff B. Hormonal abnormalities in obesity. *Acta Med Second* (Suppl.) 723:153-160. 1988.

21. Poortman J, Thijsen JHH, DeWaard F. Plasma Oestrone, Oestradiol and androstenedione levels in post-menopausal women; relation to body weight and height. *Maturitas* 3:65-71. 1981.

22. Judd HL, *et. al.* Endocrine function of the postmenopausal ovary: concentrations of androgens and estrogens in ovarian and peripheral vein blood. *J Clin Endocrinol Metab* 39:1020-1029. 1974.

23. Sitteri PK. Adipose tissue as a source of hormones. *Am J Clin Nutr* 45:277-282. 1987.

24. Schindler AE, Ebert E, Friedrich E. Conversion of androstenedione to estrone by human fat tissue. *J Clin Endocrinol Metab* 35:627-630. 1972.

25. Armstrong BK, *et. al.* Diet and reproductive hormones: a study of vegetarian and non-vegetarian postmenopausal women. *J Natl Cancer Inst* 67:761-767. 1981.

26. Shultz TD, Leklem JE. Nutrient intake and hormonal status of premenopausal vegetarian Seventh Day Adventists and premenopausal nonvegetarians. *Nutr Cancer* 4:247-259. 1983.

27. Rose DP, *et. al.* Effect of a low-fat diet on hormone levels in women with Cystic Breast Disease 1. serum steroids and Gonadotropins. *J Natl Cancer Inst* 78:623-626. 1987.

28. Prentice R, *et. al.* Dietary fat reduction and plasma estradiol concentration in healthy postmenopausal women. *J Natl Cancer Inst* 82:129-134. 1990.

29. Goldin BR, *et. al.* Effects of diet and lactobacillus acidophilus supplements on human fecal bacterial enzymes. *J Natl Cancer Inst* 64:255-261. 1980.

30. Reddy BS, Weisberger JH, Wynder EL. Fecal B-glucuronidase control by diet. *Science* 183: 416-417. 1974.

31. Adlercreutz H, Martin F. Biliary excretion and intestinal metabolism of progesterone and estrogens in man. *J Steroid Biochem* 13:231-244. 1980.

32. Bernstein L, *et. al.* Serum hormone levels in pre-menopausal Chinese women in Shanghai and white women in Los Angeles: results from two breast cancer case-control studies. *Cancer Causes and Control* 1:51-58. 1990.

33. Woods MN, *et. al.* Low-fat, high-fiber diet and serum estrone sulfate in premenopausal women. *Am J Clin Nutr* 49:1179-1183. 1989.

VARIATIONS IN ESTROGEN METABOLISM
Commentary on B. Goldin

Leon Bradlow[1]

DIETARY AND PHARMACOLOGICAL MODULATION OF ESTROGEN METABOLISM

I'd like to thank Rachel and the organizing committee for inviting me to participate in this meeting. It has been very informative thus far.

I'm going to talk about our own studies on estrogen metabolism which complement closely the results which Dr. Goldin has just described to you, and I think they explain why some of these results were observed.

This work was actually initiated almost 40 years ago when Dr. Thomas Gallagher brought me to the Sloan-Kettering Institute to start a program on the metabolism of labeled hormones in people, and it has progressed since then with a number of other investigators, most notably Barnett Zumoff, Jack Fishman, Richard Hershcopf, Karl Anderson, and John Michnovicz. The fundamental question that needs to be addressed is how do we explain why dietary alterations act to alter plasma estrogen levels. To do this we really have to look at some skeletons in the closet which have not yet been mentioned which are very critical.

Figure 1 illustrates two points that have to be kept in mind; one, that estrogens are formed by aromatization of androstenedione and testosterone which are converted to estrone and estradiol respectively. Estradiol, of course, is the principle active female hormone. Aromatization is regulated in part by fat cells in the body and by dietary intake of fat.

H. L. BRADLOW • Strang Cornell Cancer Research Laboratory, 510 E.73rd Street, New York, NY, 10021-4004.

Steroid Contraceptives and Women's Response,
Edited by R. Snow and P. Hall, Plenum Press, New York, 1994

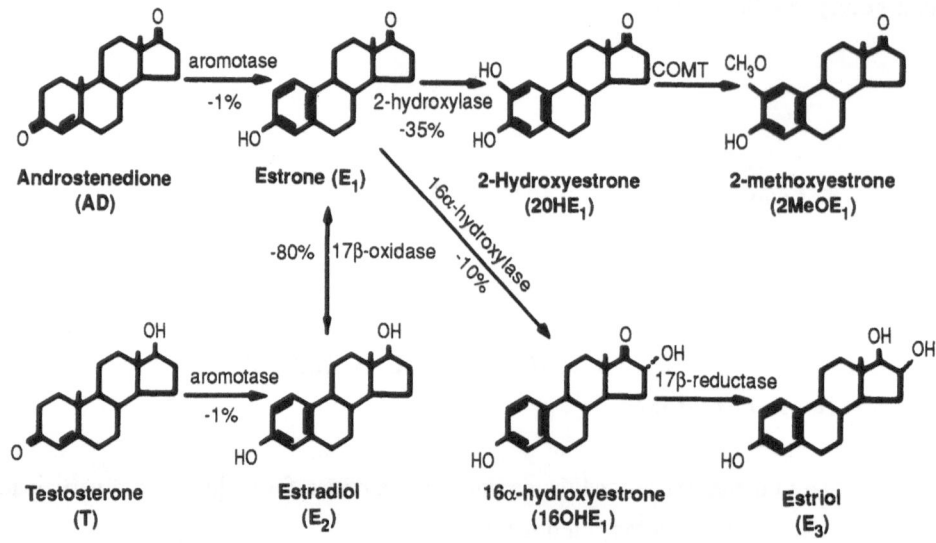

Figure 1. Biosynthesis and metabolism of estrogens in man.

Secondly, estradiol in man is initially metabolized almost exclusively by oxidation to estrone, which is then further metabolized by one of two alternative pathways and conversion to these two endpoints makes a biological difference. Estrone may either be 2-hydroxylated to yield 2-hydroxyestrone, which is a weak estrogen, essentially devoid of peripheral biological activity, or it can be 16α-hydroxylated to 16α-hydroxyestrone, which a} is in itself a potent estrogen with some interesting and remarkable properties and b} can be further metabolized to estriol, which is also a fully potent estrogen. Thus physiological events that alter the relative extent of these two pathways ultimately determines the total estrogenic response of any given amount of secreted hormone. Now, one of the critical things in postmenopausal women is that, of course, ovarian production of estradiol essentially ceases and that all of the circulating estrogen is derived by peripheral conversion from androstenedione. The major site of this pathway is adipose tissue, so that obese women make more circulating estrogen and consequently have higher estrogen levels -there is simply more estrogen available in obese women and if you reduce their weight, total estrogen production goes down.

The other thing that happens is that there are changes in the two metabolic pathways referred to above. In order to look at these pathways in a relatively efficient and simple manner, we devised a technique whereby the estradiol labeled with tritium at the metabolically reactive sites 2, 16α, and 17α is administered iv in vivo. Following metabolic transformation in the body, the total extent of each reaction is measured by the amount of tritium liberated and transferred to the body water pool. All that is necessary to do to determine the extent of any of these reactions is to determine the specific activity of the circulating water by lyophilization of blood or spot urine samples and measuring its specific activity. This value, together with the body water volume, preferably determined

Table 1. Hydroxylation of estradiol at the 2, 16α, and 17α positions in normal men and women.

Subjects	% Hydroxylation ± SD		
	2-Hydroxylation	!6α-Hydroxylation	17 Oxidation
Men	20.8 ± 7.6 (19)•	7.5 ± 1.9 (19)	62.9 ± 10.5 (20)
Premenopausal Women	38.9 ± 1.9 (6)	10.7 ± 2.1 (7)	83.8 ± 15.4 (7)
Postmenopausal Women	32.6 ± 2.6 (7)	8.5 ± 1.4 (9)	73.7 ± 18.5 (6)

( )in all cases Number of Subjects The values in the men and premenopausal women are highly significantly different $P < 0.01$.

using the bioimpedence method which Dr. Snow referred to yesterday, gives us the circulating radioactivity present as water. The product of these two values divided by the dose administered gives the extent of reaction.

As shown in Table 1, we can see what happens in men, pre-menopausal women and post-menopausal women. C_2 hydroxylation, particularly in pre-menopausal women, is a quantitatively more important pathway than 16α-hydroxylation, and of course 17-hydroxylation is much greater than either of the other two pathways. The pathways are relatively time invariant in that there is not a great deal of difference between pre- and post-menopausal subjects.

However, as illustrated in Figure 2, when we studied post-menopausal women with breast cancer, there is a considerable increase in the extent of 16α-hydroxylation in these subjects as compared with normal controls of the same age, while the extent of the other two pathways, (C-2 and C-17) were essentially unaltered. Thus the difference is confined to this one metabolic pathway, and there is very little overlap between the two populations. Now these results, of course, are in women who already had breast cancer, and one can argue that having breast cancer could alter hormone metabolism. To get around this

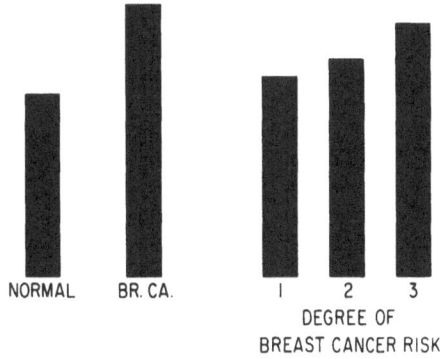

Figure 2. Metabolite 16 in women.

problem we looked at women who were at risk for breast cancer for familiar reasons ---
mother, sister, aunt, grandmother with breast cancer or because they have various kinds of
breast dysplasias that put them at increased risk. They were categorized in terms of
analytic risk and the extent of 16α-hydroxylation of estradiol measured. Women in the
upper tertile of risk had a significantly greater extent of reaction at C-16α than women in
the lower tertile of risk (14.5 vs 11.1 % P < 0.05).

As an alternative approach we have also studied the extent of 16α-hydroxylation of
estradiol in mice with different degrees of mammary tumor incidence. All of the mice
were 6-8 weeks old well before the appearance of spontaneous tumors. Those strains with
the highest incidence of tumors (90-100%) had the highest rate of 16α-hydroxylation (18-
20%) while those strains with a low rate of tumors also had a lower extent of reaction at
C-16α (3-10%). The extent of reaction is not sex linked but is inherited. If we remove
the mouse mammary tumor virus by foster nurturing from birth the extent of reaction is cut
in half (18% to 10-12%) while infecting virus free mice results in doubling the extent of
this reaction (3% to 6%).

Can these reactions be altered by dietary and drug manipulations? As Figure 3
shows, diet has a powerful effect on estrogen metabolism. On a high protein diet there
is more 2 but not 16α-hydroxylation in human volunteers on a high protein than a high
carbohydrate diet. Reaction at C-16α appears to be essentially constitutive and not readily
altered. The only dietary change which produced a significant alteration was an extreme
change in fat consumption in baboons (65% fat) which caused a significant increase in

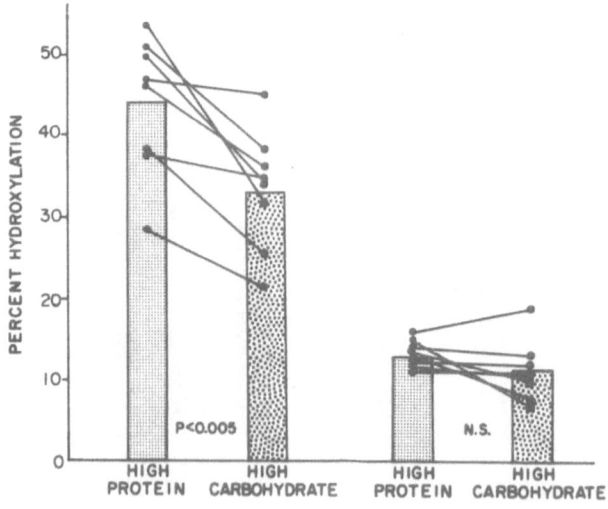

Figure 3. Effects of diet on estrogen metabolism.

Table 2. Effects of "high fiber" diet on urinary
estrogens in women.

METABOLITE OR RATIO	PRE	POST-FIBER[a]	P
[E1]	10.3 ± 2.7[b]	10.8 ± 2.7	NS
[E3]	9.3 ± 2.8	8.2 ± 3.2	NS
[2OHE1]	11.5 ± 3.2	12.8 ± 4.1	NS
[2OHE1]/[E3]	1.32 ± 0.35	1.63 ± 0.40	<0.05

[a]Every second and third day subjects (n=6) received
either 50 g cabbage or 100 g broccoli as part of a
high fiber diet.

[b]Results expressed as µg/g creatinine.

reaction at C-16α and a drop in reaction at C-2. Moderate increases in thyroxine increases
C-2 hydroxylation, while at higher levels of T-3, reaction at C-16α also decreases. In
myxedematous subjects the opposite effect is observed. In an additional study we were
also fortunate enough to obtain some urine samples from Drs. Gorbach and Goldin from
patients who had been on a low fiber diet and again on a high fiber diet which also
included cruciferous vegetables. As Table 2 shows, 2-hydroxylation increased on the high
fiber diet and 16α-hydroxylation decreased. In studies by Drs. Goldin and Gorbach a
similar change in reaction at C-16 was observed when labelled estradiol was administered
to subjects both intravenously and orally. Comparisons of Asian migrants on their home
diet and local Bostonians showed similar differences.

Body weight also has an important effect on the extent of 2 but not 16α-hydroxyla-
tion. When the extent of these reactions were compared in normal and obese subjects (both
male and female) reaction was C-2 was markedly depressed while reaction at C-16α was
unchanged (Figure 4). As a consequence there is more estrogen available to be 16α-
hydroxylated and also higher estradiol levels. In the opposite circumstance in anorectic
subjects and athletes who exercise vigorously C-2 hydroxylation is markedly increased and
there is less circulating estradiol and 16-hydroxylated products. As a consequence some
of these women become reversibly amenorrheic until they cease vigorous exercise or gain
a moderate amount of weight.

The effects of fatty acids can be demonstrated not only in the total body, but also
in target tissues for breast cancer, specifically the terminal duct lobular unit (TDLU) of the
breast, the end organ in the breast where milk is produced and where most tumors
originate. If we measure ras oncogene expression or estradiol 16α-hydroxylation in TDLU
both are stimulated by linoleic or arachidonic acids while the opposite effect is seen when
the w-3 fatty acids are added to the incubation. Benzpyrene and dimethylbenzan-thracene

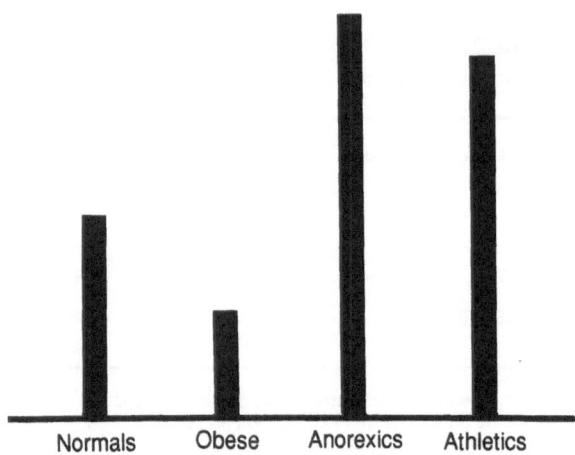

Figure 4. Metabolite 2 and body weight.

cause a similar increase in both parameters with an additive effect when linoleic and arachidonic acids were added in addition to the hydrocarbons. At the same time C-2 hydroxylation is also decreased. Other studies have shown that transfection of mammary epithelial cells with the ras or myc oncogenes or cervical cells with the herpes virus all result in an increase in reaction at C-16α and a decrease in reaction at C-2.

Drugs and minor dietary substances also proved to have potent effects on C-2 hydroxylation. Thus there is a major increase in C-2 hydroxylation in smokers as compared to nonsmokers. At the same time plasma estradiol is decreased. Both of these correlate with the observation that smoking is protective against endometrial cancer and possibly breast cancer. While this is true for estradiol as pointed out earlier it is not applicable to ethinylestradiol which is metabolized by a different member of the P450 family. A similar effect is seen with dioxins. Both in animals and in cell cultures it can be demonstrated that dioxins specifically increase the extent of 2-hydroxylation. This correlates well with the observation that women who were exposed to dioxins in a plant explosion some 15 years ago have a decreased incidence of breast cancer (RR ~ 50%). The opposite effect is observed in volunteers treated with Tagamet. It had been observed for some time that patients on this β-blocker often develop gynecomastia and other signs of secondary feminization. Examination of 2-hydroxylation showed that this reaction is inhibited resulting in increased plasma estradiol and urinary 16-hydroxylated metabolites even though the extent of 16-hydroxylation measured radiometrically is unchanged. The reaction is quite specific and is not seen with Zantac. Because the body is an unsaturated system in the extent of one of these pathways can result in the product of the other reaction event though the extent of the other pathway is unaltered.

The question which is raised by these findings is whether there is a safe way of increasing estradiol 2-hydroxylation to exert this protective effect. This proved to be very simple to achieve. A prototype compound, indole-3-carbinol, found in cabbage, broccoli, brussel sprouts, and other cruciferous vegetables proved to very potent in inducing estradiol-2-hydroxylation. Studies in cell culture, animals, and people have all demonstrated that indole-3-carbinol specifically induces 2 but not 16α-hydroxylation. The increase in women amount to 50% or greater after only one week on the compound. In animal studies it was possible to demonstrate that I-3-C specifically induced P45021A2 and 1A1 and not any others. Finally, in feeding studies in C3H/OuJ mice maintained on 2000 ppm of I-3-C diet as compared to animals on the control diet, the incidence of tumors was markedly decreased.

BODY FAT, FAT TOPOGRAPHY AND ENDOGENOUS STEROID DYNAMICS

Rachel Snow[1]

Dr. Garza-Flores and his colleagues have presented evidence of systematic differences in the pharmacokinetic profiles of depo medroxyprogesterone acetate (DMPA) and norethisterone enanthate (NET-EN) between Mexican and Thai women. Comparing these data with similar studies conducted in other regions (cited in the same paper), it appears that the pharmacokinetic parameter displaying the greatest variability between populations is the C_{max}, the maximum concentration observed; it is the mean C_{max} observed with DMPA among Thai women that is the most dramatic outlier.

These (and other such data from Drs. Goldzeiher, Fotherby and Sang) prompt the question, what is region?   A complex interplay of environmental and inherited factors. Two such factors, diet and body composition, warrant closer attention with respect to variability in contraceptive steroid metabolism, because both diet and body composition are associated with variability in *endogenous* steroid metabolism. As several chapters in this volume look exclusively at the role of diet and steroid dynamics (see chapters by Bradlow, Goldin and Longcope in this volume), I devote this chapter to a brief review of associations between body composition and the dynamics of endogenous reproductive steroids.

Body Composition - What is it?

There are several dimensions to body composition, and it may be helpful to address them categorically: the first and crudest dimension is the absolute mass of the body,

R. SNOW • Department of Population and International Health, Harvard School of Public Health, 665 Huntington Avenue, Boston, MA 02115

Steroid Contraceptives and Women's Response,
Edited by R. Snow and P. Hall, Plenum Press, New York, 1994

expressed as weight alone or as weight per unit standing height (body mass index, Quetelet's index or ponderal index); a second is the composition of the body, or the relative proportions of specific bodily components, e.g. fat, lean mass, water, or bone; finally, there is the topography of the body, or the relative distribution of mass or fat across different regions of the body, often evaluated by the ratio of waist/hip circumference.

Body composition obviously interacts with nutritional intake, and both are subject to inherited and environmental influences. Intake is largely environmental, and directly affects body composition, but many dimensions of body composition are partially inherited, including height potential, a predilection to leanness or fatness, the density of bone, and the relative distribution of fat and lean mass over different regions of the body, i.e. fat topography.

Relative fatness and fat topography are each associated with blood concentrations of certain reproductive steroids, with the concentration of sex hormone-binding globulin (SHBG) (and hence the bio-availability of select steroids), and with the liver metabolism of estrogens.

Body weight, body composition and sex steroids

In the early the 1970's attention was drawn to the phenomena of secondary amenorrhea in underweight women (Frisch and McArthur 1974). Anorexia nervosa garnered increased clinical attention (Ross Conference 1983), and several investigators reported that anorectics had a blunted release of gonadotropic releasing hormone (GnRH). Low and irregular GnRH pulses lead to low plasma gonadotropins and suspended follicular development. Re-feeding and restoration of normal body weight in these women was associated with full (and progressive) recovery of GnRH and gonadotropin release, and a subsequent return of normal menses (Vigersky et al. 1977; Nillius 1983).

As among women with low weight for height, women of low relative fatness have low gonadotropins, follicular activity suspended in early follicular phase, and consequent low concentrations of estrone, estradiol and progesterone (Loughlin 1985; Cumming 1987).

Such observations gave rise to speculation regarding mechanisms that may be responsible for the association between low weight and menstrual function. Frisch (1990) has hypothesized that the cessation of GnRH among underweight women occurs because such women have dipped below a critical fat threshold essential for maintaining adequate levels of sex steroids. Studies of body composition among amenorrheic and eumenorrheic women have generally supported an association between low body fat and menstrual disturbance (Cumming 1987), but the interaction of body fat, sex steroid concentrations and GnRH regulation is multi-faceted, and remains largely unexplained.

If low fatness is associated with aberrations in the GnRH, gonadotropin-ovarian axis, what other co-variates of low body fat are associated with either GnRH function, or the levels and bioactivity of the ovarian hormones whose feedback modulates GnRH regulation? Two such factors will be discussed here: sex hormone-binding globulin (SHBG) and estrogen metabolism.

Sex Hormone-Binding Globulin

SHBG is a principal binding protein for estradiol, and shifts in SHBG concentration affect the proportion of free estrogen, or that fraction available for bioactivity. SHBG not only varies with body fatness but responds rapidly to changes in dietary intake. In anorexia nervosa SHBG levels are high, and SHBG levels decline with re-feeding (Estour et al 1986); in two weeks of dieting among normal weight women, SHBG levels increased two-fold (Kiddy et al 1989); and SHBG levels tend to be higher in leaner women (Kurzer and Calloway 1987). Conversely, in obese women SHBG levels are low (O'Dea et al 1979; Evans et al 1983). Peiris (1989) showed a strong inverse correlation between fat mass (by underwater weighing) and concentration of SHBG.

Estrogen Metabolism

The association between low body fat and menstrual dysfunction may also be due to fatness-related changes in estrogen metabolism. Fishman (Fishman et al. 1975; Schneider et al. 1983) found an association between body weight and the pathway of estrogen metabolism that may contribute to menstrual dysfunction among lean individuals. There are two distinct pathways of estrogen metabolism, 16 alpha-hydroxylation of estrone leading eventually to estriol, and 2-hydroxylation of estrone leading to the catechol metabolites 2-hydroxyestrone and 2-methoxyestrone. The uterotrophic activities of the products of these two pathways are markedly different: 16 alpha-hydroxylation allows retention of estradiol's uterotrophic potency, whereas 2-hydroxylation produces estrogens with greatly reduced biological activity.

Using radiometric techniques to assess the percent of estradiol metabolized by each of these two pathways, Fishman found that obese individuals have a dramatically diminished percent of non-potent catechol formation when compared to normal individuals (Schneider et al 1983). The results of previous urinalyses of estrogen metabolites in obese, normal and anorectic women had shown that very lean, anorectic women had significantly greater catechol metabolites (Fishman et al 1975).

More recently, Frisch et al (1993) has demonstrated that among women of equal weight but different relative fatness, total fat as a percent of total area (estimated by magnetic resonance imaging) is inversely associated with the extent of estrogen 2-hydroxylase activity.

Fat Topography

If you separate obese individuals by their waist to hip ratios, there are two general types: people who carry their excess body fat in the upper trunk, particularly above the umbilicus (android obesity), and individuals who carry their extra body fat below the level of the hips (gynoid obesity). Women with android obesity are prone to metabolic pathologies, including an elevated risk of diabetes (Vague et al 1985; Krotliewski et al 1983), hypertension, hyperlipidemia and myocardial infarction (Larsson et al 1984; Lapidus et al 1984), and they are more likely to have decreased SHBG concentrations than gynoid, but equally obese, individuals (Kissebah 1982; Evans et al 1983).

Obesity is often associated with insulin resistance, and it is now apparent that upper trunk obesity is associated with higher plasma insulin and greater insulin resistance than lower trunk obesity (Bjorntorp 1988, Kissebah 1982; Peiris et al 1987).

Recently, there has been growing evidence that the observed association between fatness and SHBG may be attributable to changes in insulin concentration, whereby insulin actually regulates changes in SHBG production. In vitro studies using human hepatoma cell line indicate that insulin inhibits SHBG production by the human hepatoma (Hep G2) cell line (Plymate et al 1988). (Hepatocytes are the main, if not the only, cells producing SHBG.) In a controlled study of obese women with polycystic ovaries (PCO) in which ovarian steroid production was suppressed by a GnRH-agonist, complete suppression of insulin release by diazoxide treatment led to a 32 percent increase in serum SHBG after 10 days (Nestler et al 1991). These data give evidence that hyper-insulinemia directly reduces SHBG levels, independent of effects by ovarian steroid concentrations.

It appears, therefore, that the association between body composition, fat topography and SHBG may be mediated by insulin dynamics. Android obesity is associated with lower metabolic clearance (MCR) of insulin, increased plasma insulin post-challenge (more insulin resistance), and lower SHBG (Peiris 1987). Fasting insulin is negatively correlated with SHBG levels (Peiris et al 1987), and the correlation of adiposity with SHBG is significantly reduced when adjusted for hyper-insulinemia (Peiris et al 1989), suggesting that the relationship between SHBG and insulin may even be independent of adiposity. Peiris (1989) speculates that portal hyper-insulinemia of obesity exposes the liver to high insulin, decreasing SHBG. But why more portal hyper-insulinemia?

Abdominal fat is highly lipolytic, relative to fats from other depots in the body. Therefore, an excess of body fat in the upper trunk, particularly in the abdominal region, may be associated with greater hepatic exposure to free fatty acids (Bjorntorp 1988).

Fat Topography and Ovarian Function

Having outlined the above, what is the evidence for any clinical association between fat topography and ovarian function? Lean women with more internal fat in abdominal regions show greater evidence of follicular activity, compared to equally lean women with low fat depots in these regions (Frisch et al. 1993). Given Bjorntorp's (1988) report that women with intact ovarian function have *more* lipolysis in the abdominal region than menopausal women, and that the drop in abdominal lipolysis is proportional to the drop in estradiol, the observed association of intact ovarian function and *more* abdominal fat deposition is particularly provocative. While Bjorntorp was citing his observation to illustrate the impact of sex steroids on body fat topography, the Frisch data open the possibility of some reverse causality.

Future Investigations

There are obviously many unanswered questions regarding the interactions of body composition, fat topography and sex steroids. This review of the principal literature in this area has been provided to introduce the contraceptive researcher to clinical observations of potential overlapping interest.

With respect to future directions for research, I offer several suggestions. First, as outlined earlier in this volume (see Snow and Wilson), the principle question for family planning researchers is whether the observed variability in contraceptive steroid dynamics has any bearing on women's actual experience of these drugs: are varying blood levels, and different pharmacokinetic profiles among different women associated with varying degrees of ovarian suppression, or varying bleeding responses to the methods, or with any other significant side-effects?

In the event that blood concentrations and pharmacokinetic profiles are associated with contraceptive side-effects and other clinical outcomes, then it is becomes worthwhile to investigate the biologic basis of variability in women's metabolic response to these steroids. At that point, factors known to affect endogenous steroids become signposts for future investigations, and measures of body weight, body composition, fat topography and diet are warranted.

To date, researchers, have been content to characterize build as height and weight, and sometimes as weight alone, failing to disaggregate fat from lean mass or heavy from light bones. Consequently, all chance of a physiological analysis of pharmacokinetic variability has been lost (Tanner 1989).

Simple measures of body composition and waist-hip ratios are not expensive or laborious measures to conduct on contraceptive clients, especially if clients are already taking part in research protocols such as clinical trials. For example, bio-electrical impedance analysis (BIA) offers a simple, non-invasive, and reproducible estimate of body composition (Lukaski et al 1985). The machine required for BIA measures is cheap and portable; clients can remain clothed and need only to lie prone for 1-2 minutes while the measurement is made. Estimates of body composition from BIA are highly correlated with direct measures of density and total body water (Lukaski et al 1985), far more so than estimates made from caliper testing or anthropometric measures.

In studies wherein blood concentrations of steroids are being monitored in relation to pharmacodynamic parameters, it may be worthwhile to examine more closely the interaction of contraceptive steroids, binding proteins, and the free/bound fraction of endogenous steroids. Underlying variability in SHBG synthesis (due to fat topography and insulin dynamics) may, in fact, mediate the impact of contraceptive steroids on ovarian function, or other significant pharmacodynamic parameters.

References

Björntorp P. Fat cell distribution and metabolism. *Annals of the New York Academy of Sciences* 66-72. 1988.

Estour B, Pugeat M, *et al*, Sex hormone binding globulin in women with anorexia nervosa. *Clin. Endocrin.* 24:571-576, 1986.

Evans DJ, Hoffman R, et al., Relationship of androgenic activity to   body fat topography, fat cell morphology, and metabolic aberrations in premenopausal women. *J. Clin. Endocr. Metab.* 57:304-310, 1983.

Fishman J, Boyar R, Hellman L. Influence of body weight on estradiol metabolism in young women. *J Clin Endocrinol Metabo* 41:989-991. 1975.

Frisch RE. Body fat, menarche, fitness and fertility. *Progress in Reproductive Biology and Medicine* 14:1-26. 1990.

Frisch RE, *et al*. Magnetic resonance imaging of overall and regional body fat, estrogen metabolism, and ovulation of athletes compared to controls. *J Clin Endocrinol Metab* 77(2):471-477. 1993.

Frisch RE and McArthur JW. Menstrual cycles: fatness as a determinant of minimum body weight for height necessary for their maintenance or onset. *Science* 185:949-951, 1974.

Kiddy D, *et al*, Diet-induced changes in sex hormone binding globulin and free testosterone in women with normal or polycystic ovaries: correlation with insulin and insulin-like growth factor-I. *Clin Endocrin.* 31:757-763, 1989.

Kissebah A, *et al*. Relation of body fat distribution to metabolic complications of obesity. *J Clin Endocrinol Metab* 54(2):254-260. 1982.

Krotkiewski M, *et al*, Impact of obesity on metabolism in men and women. Importance of regional adipose tissue distribution. *J Clin Invest* 72:1150-1162, 1983.

Kurzer M, Calloway DH, Effects of energy deprivation on sex hormone pattern in healthy menstruating women. *Am J Physiol.* 251(E14):483-488, 1987.

Lapidus L, *et al*, Distribution of adipose tissue and risk of   cardiovascular disease and death: a 12 year follow up of participants in the population study of women in Gothenburg, Sweden. *Br Med J.* 289:1257-1261, 1984.

Larsson B, *et al*, Abdominal adipose tissue distribution, obesity and risk of cardiovascular disease: 13 year follow up of participants in the study of men born in 1913. *Br Med J.* 288:1401-1404, 1984.

Lukaski HC, *et al*, Assessment of fat-free mass using bioelectrical impedance measurement of the human body. *Am J Clin Nutrit.* 41:810-817, 1985.

Nillius SJ. Weight and the menstrual cycle. In: *Understanding Anorexia Nervosa and Bulimia.* Report of The Fourth Ross Conference on Medical Research. Columbus, OH: Ross Laboratories. 1983.

Nestler JE, *et al*, A direct effect of hyperinsulinemia on serum sex hormone-binding globulin levels in obese women with the polycystic ovary syndrome. *J Clin Endocr Metab* 72:83-89, 1991.

O'Dea J, *et al*, Effect of dietary weight loss on sex steroid   binding, sex steroids, and gonadotropins in obese postmenopausal women. *J Lab Clin Med* 93:1004-1008, 1979.

Peiris AN, *et al*. The relationship of insulin to sex hormone-   binding globulin: Role of adiposity. *Fertility and Sterility*   52(1):69-72. 1989.

Peiris AN, Struve MF, Kissebah AH. Relationship of body fat distribution to the metabolic clearance of insulin in premenopausal women. *Int J Obes* 11:581-589. 1987.

Ross Conference on Medical Research. *Understanding Anorexia Nervosa and Bulimia.* Report of The Fourth Ross Conference on Medical Research. Columbus, OH: Ross Laboratories. 1983.

Snow R, Barbieri R, Frisch R. Estrogen 2-hydroxylase oxidation and menstrual function among elite oarswomen. *J Clin Endocrinol Metab* 69(2):369-376. 1989.

Svedberg J, *et al*. Fatty acids in the portal vein of the rat regulate hepatic insulin clearance. *J Clin Invest* 88:2054-2058. 1991.

Tanner, Human growth and constitution in *Human biology* Harrison, Tanner, Pilcan and Baker. *Oxford Science* 3rd Edition, 1987.

Vigersky R, *et al*. Hypothalamic Dysfunction in Secondary Amenorrhea Associated with Simple Weight Loss. *N Engl J Med* 297(21):1141-1145. 1977.

Vague J, *et al*, in *Metabolic complications to human obesities. Excerpta Medica Amsterdam* Vague et al, Eds:3-12.

STEROID CONTRACEPTIVES AND WOMEN'S RESPONSE: PROGRAM CONSIDERATIONS FROM THE QUALITY OF CARE PERSPECTIVE

Joan Kaufman[1]

1. INTRODUCTION

Women vary not only in their biological response to contraceptives, but also in their assessments of acceptability of method characteristics and the services for providing them. Method discontinuation, while often mainly attributable to unacceptable side effects, may also be caused by unacceptability of the method's mode of action, route of administration, or poor experience with the service system that provides the method. As biomedical researchers strive to understand biological variability in method response and to help fine tune methods to minimize potential side effects, family planning service systems may simultaneously introduce mechanisms for increasing contraceptive acceptance and continuation by improving routine program operations that increase client satisfaction with methods.

Variability in women's response to contraceptive methods, steroidal or other, necessitates that service systems which provide those contraceptives to women must be able to service both different individuals and changes in the same individual, biologically and with respect to reproductive intentions. The quality of care framework for family planning services, developed by Bruce (1989), provides a starting point for considering the service requirements of programs necessary for increasing the acceptability of contraceptive

J. KAUFMAN • Department of Population and International Health, Harvard School of Public Health, 665 Huntington Avenue, Boston, MA 02115

services for women who vary in reproductive intentions, cultural belief systems, environments, health risks, and biological responses to contraceptives.

Quality of care refers to six main features of services which contribute to safe, effective, and satisfied contraceptive use by women. An assessment of the quality of services is a client oriented process which focuses attention to the interface of the client with the service delivery system. Six measurable aspects of quality relate to choice of methods, information given to clients, technical competence of service providers, continuity and follow up, interpersonal relations, and appropriate constellation of services (Bruce, 1989). These quality characteristics are highly inter-related. Together, they constitute the capacity of a family planning service system to respond to individual variability in both biological and behavioral responses to contraceptive methods, their delivery systems, and side effects.

Between 60-70 million women worldwide are current oral pill users (United Nations, 1987), another 355,000 women use Norplant[R] (WHO, 1990), and by the early 1980's an estimated 2.3 million women were using some type of contraceptive injection such as depo-provera (DMPA) (Liskin, 1983). This paper will discuss the evidence presented at this meeting about women's varying biological response to these most commonly used methods of steroidal contraception, discuss issues of cultural and social acceptability of steroid induced side effects of these methods, and suggest ways in which programs may respond to these variations so as to promote safe, effective, and satisfied use of these methods. Quality requirements for family planning services will be reviewed with specific reference to the provision of the pill, injectable, and Norplant[R] in developing country settings.

2. STEROID ACCEPTABILITY: BIOLOGICAL ISSUES

Several of the papers presented at this conference point out that steroid metabolism may be influenced by a variety of factors and may vary not only between women, but also within the same woman. There is evidence to suggest that different ethnic groups may metabolize steroids at different rates (Goldzieher, 1990). And, these already variable processes may be further influenced by interactions with certain drugs commonly used in developing countries, such as antibiotics and anti-tuberculosis drugs (Back, 1990) as well as by nutritional and energy expenditure factors. Behavioral and demographic factors, such as smoking and age of the woman also influence the effect of steroids on the body.

This variability in steroid metabolism may be linked to the occurrence of side effects in both pill, injectable and Norplant[R] users. While there is wide variability in bleeding patterns among women worldwide (Belsey, 1990), disruptions in menstruation occur for all three methods and are the most commonly cited reason for method discontinuation of the injectable and Norplant[R]. In 1977, Kreager has noted that discontinuation of the pill due to minor side effects accounted for about one third of all reasons for discontinuation with total mean discontinuation rates for the pill as high as 30% worldwide. For pill users, these side effects include breakthrough bleeding, reductions in menstrual blood loss, urinary tract infections, vaginal discharge and irritation, chloasma, weight change, nausea, depression, headache, mastalgia. Different formulations of the oral pill, combined, estrogen, or progestogen only may elicit different responses from both the same and different women. For injectable users, changes in menstrual bleeding patterns, particularly spotting and irregular bleeding are main side effects, although many of the same complaints from pill use are reported. In early studies, one fourth to one half of all women who discontinued using depo-provera did so because of disruptions of the menstrual cycle (Rinehart 1975). For Norplant[R] the most commonly cited side effect is irregular bleeding, although weight change, dizziness, fatigue, and headache have also been reported (Zimmerman, 1990; Sivin, 1988). These unacceptable side effects are also the most frequently cited reason for removal of Norplant[R] (Sivin, 1988; Zimmerman, 1990).

Serious health risks of hormonal methods, while uncommon, have been recognized for quite some time. These include thromboembolism, thrombophlebitis, stroke, subarachnoid hemorrhage, heart attack, hypertension, benign liver tumors, gallstones. Some newer serious risks, such as a possible increased risk of HIV infection by users of oral pills and long acting steroidal methods, are also suspected and have widespread implications for high prevalence HIV areas (Hunter and Mati, 1990).

3. STEROID ACCEPTABILITY: CULTURAL AND PERSONAL ISSUES

Side effects of steroid contraceptives usually constitute small or insignificant health risks. Nevertheless these side effects may be socially or culturally unacceptable with respect to indigenous belief systems and lifestyles. Disruptions in menstrual bleeding patterns appear to be the most culturally unacceptable. Women in many societies have strong beliefs about the timing, quantity, predictability, and significance of menstruation.

These beliefs about menstruation may make changes in bleeding associated with pills, injectable, and Norplant[R] unacceptable. For example, in traditional Islamic and Orthodox Jewish societies, menstrual blood is considered unclean, and men-struating women forbidden from attending the mosque or temple, food preparation, and sexual relations (Shain, 1980). Other menstrual taboos in other societies include proscriptions against visiting religious places, cooking, bathing, washing clothes and hair, visiting pregnant or newly delivered women (Nag, 1984). Some cultures regard menstruation as a monthly purge which cleans the blood and thus find methods which cause amenorrhea, such as the injectable, unacceptable (Shain, 1980). In China, both too much and too little menstrual blood loss are suspected of causing an imbalance in the "yin yang" properties of the body and oral contraceptive pills are thought to cause "hot" symptoms which negatively affect health (Ngin, 1985; Chu, 1977).   In these societies, contraceptive induced amenorrhea, spotting, breakthrough bleeding, or reduction in menstrual blood loss may pose serious social or psychological problems for women.

Other health beliefs may affect the acceptability of a method in a specific setting. For example, in Indonesia, Norplant[R] acceptability has been affected by confusion with a traditional custom of implanting precious metal into the body thought to impart strength and beauty but widely associated with prostitution (Zimmerman, 1990).

4. STEROID ACCEPTABILITY: SERVICE RELATED ISSUES

Service systems for providing contraceptives to women in developing countries have expanded greatly in the last several decades. Previously, services were provided mainly by physicians in clinical settings. Today, in much of the world, private outlets, community based distribution and social marketing programs have all provided mechanisms for greatly expanding the reach of contraceptive services to previously under served or overlooked segments of the population. With this expansion has come a complexity both in type and training of personnel providing those contraceptives and in the characteristics of facilities where they are obtained. Often, contraceptive service delivery is detached from primary health care programs which attend to other reproductive health concerns, such as maternal and child health, treatment of sexually transmitted diseases, or the provision of basic health care and drugs.

To add to this complexity, women often use multiple service sites to obtain contraceptive services, especially if they switch to methods unavailable at their current outlet, or discover more convenient sources of resupply after initial method acceptance. This fragmentation in sources and utilization create challenges for program planners to insure that methods are being used safely and effectively by women who accept them. Because of this fragmentation, attention to the quality of services delivered at the point of delivery (clinic, pharmacy, or by a field worker) is important. An analysis of quality factors as they relate to the delivery of steroidal contraception may help focus attention on specific program requirements.

5. METHOD CHOICE

Each method of contraception has its own attributes with regard to effectiveness, reversibility, side effects, convenience of use and cosmetic appeal. Couples assess these characteristics in light of their own specific needs and preferences at points in their reproductive life cycles. For this reason the provision of a choice of contraceptives improves more widespread acceptance and use for populations composed of different types of potential users, and increases the likelihood of method switching instead of discontinuation when method specific problems arise. Pariani (1990) demonstrated in Indonesia that 90% of women who were granted their first choice of methods were still using that method after one year, compared to only 28% who's first choice was denied and Jain has shown that addition of method choices to a program can be expected to contribute significantly to overall contraceptive prevalence (Jain, 1989).

Oral pills, injectable, and Norplant[R], while all steroidal methods of contraception, require different service approaches to make them available to users. In many countries, oral pills may be easily obtained through community based distribution programs and through pharmacies. Yet, because these outlets require constant supply, they are sensitive to shortcomings in logistics systems that resupply fieldworkers and outlets. These logistic systems constraints may limit short term method choice by creating gaps in supplies or by switches in brand names and formulations. The quality of supplies, including evenness of doses may also be compromised by poor manufacturing, storage, and shipping. Moreover, fieldworkers and distributors may vary widely in training, and thus in their ability to screen potential acceptors for contraindications, or manage side effects. Injectable contraceptives delivered at three month intervals, usually require trained personnel to administer the injection, but have obviated the need for frequent return visits to clinics. But injectables are subject to similar logistic system constraints as pills in the requirement for resupply, proper storage and handling. Norplant[R] has provided a long acting, low maintenance alternative to sterilization for women desiring long periods of protection and is relatively free of problems resulting from the need for constant resupply. But provision of the method requires physicians skilled in surgical techniques for both inserting and removing the implants. The pill offers the advantage of being self administered and easily discontinued by women who wish to stop or change the method, and both injectable and Norplant[R] may appeal to women who seek low contraceptive visibility and user involvement.

6. INFORMATION GIVEN TO CLIENTS

The provision of information to contraceptive method acceptors about potential side effects and problems resulting from contraceptive use contributes to both contraceptive continuation and client satisfaction. Whelan (1974) showed that instructions to patients are an important determinant of compliance with contraceptive regimens, especially those with side effects. Studies have shown that when providers do not provide thorough information on potential side effects for fear that clients will not accept certain methods if told in advance of possible problems, the plan may backfire and result in discontinuation and the exaggeration of rumors and fears (Keller, 1973).

Recent evidence from a four country study points out the importance of information provision for Norplant[R] users and family planning service providers. Focus group discussions revealed a multitude of erroneous beliefs about the ill effects of Norplant[R] on health, ranging from cancer and sterility to the fear that acceptors are required to provide large amounts of blood at clinics. These fears were closely tied to indigenous health beliefs in each country (Zimmerman, 1990). The same study revealed that service providers themselves had low levels of accurate information about Norplant[R], probably contributing to the low knowledge levels among potential acceptors. Reassurance about the safety of Norplant[R] as well as explanation that mid cycle bleeding was not real menstrual bleeding, and thus should not inhibit praying by Moslem women, increased the acceptability of Norplant[R] among groups of women who might otherwise have rejected the method (Zimmerman, 1990). Another study in Sri Lanka reported that counseling about menstrual disruptions assured potential acceptors that these changes were not associated with ill health, and resulted in low levels of discontinuation among users, despite complaints (Basnayake, 1988).

Information to users also entails providing information on warning signs of serious health risks, and where and when to return for follow up to insure safety in use. The biological evidence of drug interactions with pill use suggests that this information should also include clear instructions to women on the proper use of pills in conjunction with some antibiotics and anti-tuberculosis drugs.

7. TECHNICAL COMPETENCE OF SERVICE PROVIDERS

The provision of orals, injectables, and Norplant[R] each depend on different types of service providers with varying degrees of training and skill, yet all three entail similar contraindications to use and potential risks and side effects, and therefore require similar screening. The evidence of biological variability in women's response to steroids reaffirms the critical importance of both community and individual risk assessments by service personnel who provide steroidal methods to women in developing countries.

Women in developing countries are often provided contraceptives without thorough screening for these risks. Practically, medical and laboratory facilities and personnel for screening are usually unavailable where services are needed the most. Moreover, the potential short term impact of increased access to contraception on women' health (through reduced childbearing and spaced pregnancies) afforded by community based programs, often overrides concern about possible long term health risks. Nevertheless, standard procedures for improving screening without limiting access can be instituted. These many include such routine procedures as taking oral histories about risks (symptoms of STDs, hypertension, hepatitis) or conducting simple low technology practices, such as blood pressure reading. A disturbing study of low income women in Rio de Janeiro, Brazil revealed that 17 percent of oral pill users were over age 35 and 9 percent were over age 40. Of these women, 40 percent were cigarette smokers ** risk factors associated with the use of oral contraceptives. Moreover, fifty percent of these pill users reported positively to five leading health problems considered to be contraindications to pill use: varicose veins, hypertension, renal disorders, heart disease, and diabetes (Costa, 1988). Eighty five percent of these women obtained the pills from pharmacies, without a doctor's prescription. This evidence might imply that non medical distribution of orals is unsafe. Yet in direct contrast to the Rio example is the evidence that community based distribution workers in developing countries can safely provide orals, after receiving targeted training, and by using simple checklists (Atkinson et al, 1974), a strategy which is currently being promoted for both pharmacists and other private non medical providers of orals.

The skill of the provider in performing clinical tasks and the maintenance of stringent asepsis in surgical procedures may directly affect method acceptability of Norplant[R] by women. Infection and pain at the insertion site has been reported in several countries (Sivin, 1988), and skill at removal of the implants is critical to minimize scarring, a fear expressed by many potential acceptors.

The biological evidence suggesting that drug interactions, nutritional, and ethnic characteristics of populations may all affect the metabolism of steroidal methods together with evidence of medical risks to use, suggests the need for family planning program planners to make community based assessments of risks when providing steroidal based methods of contraception. These risk assessments can serve as guides to providers in a community when screening potential method acceptors. Such assessments can look at dietary intake of fat, indoles, or other nutrients suspected of affecting steroid metabolism, typical body weights and fat for women in a community, disease prevalence (hepatitis, malaria, tuberculosis, hypertension, HIV infection, chlamydia), and patterns of use of wide spectrum antibiotics or anti-tuberculosis drugs.

8. INTERPERSONAL RELATIONS

Provider attitudes can be a serious barrier or contributor to both acceptance and continuation of methods. These attitudes refer to such attributes as respect for clients and sensitivity to client concerns. Counseling skills which affect the ability to communicate necessary information between client and provider are a component of interpersonal relations. Simmons reported that in Bangladesh where female workers act as allies to the women they serve, they help to empower women to use and continue contraceptives (Simmons et al, 1988). Schrimshaw (1976) showed that in Ecuador, a country which places a high value on modesty and where sexual matters are rarely discussed, insensitivity to sexual modesty by clinic workers discouraged use of the family planning clinic by women. This insensitivity was characterized by a lack of privacy during examination, excessive and irrelevant questioning by providers who returned very little information, and rudeness and abruptness by receptionists. Mernissi (1975) reported the negative effect of provider's "degrading and dehumanizing" treatment on women's desire to return for family planning services in urban Morocco.

Mundigo reported that in Honduras, a typical two hour waiting time for resupply of pill users resulted in a drop out rate of 40% (Mundigo, 1973). Another study (Phillips, 1986) reported that workers trained in better interpersonal communication skills in Bangladesh achieved greater credibility among clients and positively affected contraceptive behavior. The importance of ascertaining fears about potential bleeding disruptions and then providing counseling to Norplant[R] acceptors has been shown to increase the acceptability of the method in Indonesia and other settings (Zimmerman, 1990; Basnayake, 1988).

9. FOLLOW UP AND CONTINUITY

Proper follow up for contraceptive acceptors includes medical checkup to check for adverse health effects from use, referral and treatment of side effects, encouraging clients to switch if problems persist or are unacceptable, determining satisfaction with use, as well as ensuring that clients are using methods correctly. Follow up mechanisms to insure safe and satisfied use of hormonal methods are particularly important given the potential for intra-individual variability and the occurrence of seasonal nutritional and disease shifts in many developing countries. Follow up may include not only health checks for contraindi-cations to use, such as changes in blood pressure, smoking status, disease status, use of interacting drugs, or dietary and body weight changes, but also assessments of the acceptability of changes in bleeding that may have occurred from steroid use and any counseling or method switching that may be necessitated by these changes.

10. APPROPRIATE CONSTELLATION OF SERVICES

The fragmentation of service delivery modes for contraceptives in many developing countries, and the separation of those services from other health services have created particular challenges for safeguarding women's health in the use of contraception. Different environments may have different health problems which may contribute to increased risk of contraceptive use by women who live there. Wasserheit (1989) reported on a previously unrecognized problem of lower reproductive tract infections among users of family planning services in rural Bangladesh which placed these women at a higher risk of ascending upper tract infections with methods of family planning involving trans-cervical procedures, such as the IUD. Recent evidence suggests that oral pill users may be at higher risk of HIV infection (Hunter and Mati, 1990). The study cited earlier from Brazil revealed that a substantial proportion of oral pill users among low income women were breast-feeding (Costa, 1988). Back (1981) has reported on problem of interactions between the oral pill and some anti-tuberculosis drugs and antibiotics, suggesting the importance of coordinating essential drug programs and family planning programs at the national and local levels, so as to provide adequate information to local practitioners on possible drug interactions associated with steroidal methods of contraception. All these studies suggest the importance of formalizing critical linkages between programs that provide health services in developing countries, so as to insure that users of steroidal methods are not placing themselves at increased risk of other health problems.

11. CONCLUSION

Acceptability of contraceptives goes beyond biological tolerance. It also includes acceptability of method induced side effects, as well a women's experience using the service system that provides those methods. Increasing contraceptive use and continuation are often stated goals of family planning programs. Yet improving contraceptive use should be viewed as part of an overall strategy to empower and improve the health of women in developing countries. Provision of contraception must be guided by considerations of women's health and satisfaction and should derive from community based assessments of the realities of these women's lives in the settings where they live: the potential effect of method induced side effects on their work capacity, social and religious relations, and their overall health status. Contraceptive acceptability research should be directed at "modify(ing) technology and programs to fit people, rather than modifying people to fit technology and programs" (Marshall, 1977).

The rapid expansion of family planning services in the last decades through alternative distribution strategies has created challenges for maintaining adequate levels of safety and informed choice. The result is that even for the known risks of contraceptive use, many service programs provide uneven screening, information, and follow up to women. The evidence presented at this workshop suggests some additional possible risks of use not currently addressed in training programs for service providers or in the development of screening and information protocols for programs, such as possible drug interactions and nutrition risks. This reaffirms the critical need to keep research closely linked to routine program operations, such as training programs for service providers, and to regularly assess the environments in which family planning services are offered, based on new information as it becomes available.

References

1.   Atkinson L, *et. al.* Oral Contraceptives: considerations of safety in nonclinical distribution. *Studies in Family Planning* 5(8):242. August 1974.
2.   Back D, *et. al.* Interindividual Variation and Drug Interactions with Hormonal Steroid Contraceptives. *Drugs* 21:46-61. 1981.
3.   Basnayake S, Thapa S, Balogh SA. Evaluation of Safety, Efficacy, and Acceptability of Norplant^R Implants in Sri Lanka. *Studies in Family Planning* 19:1. pp. 39-47.
4.   Belsey E. Regional and Individual Variation in Bleeding Patterns Associated with Steroid Contraception. Paper prepared for workshop on "Steroid Contraceptives and Women's Response", Exeter, New Hampshire, 21-24 October, 1990.
5.   Bruce J. *Fundamental Elements of the Quality of Care: A Simple Framework.* May 1989. (Population Council. Programs Division. Working paper;1.)
6.   Chu CM-Y. Menstrual Beliefs and Practices of Chinese Women. Paper Presented at California Regional Seminar in Chinese Studies, Center for Chinese Studies, Berkeley, California, February 11-12, 1977.
7.   Goldzieher JW. Pharmacology of the Ethynyl Estrogens in Various Countries. Draft of paper to be presented at workshop on "Steroid Contraceptives and Women's Response", Exeter, New Hampshire 21-24 October, 1990.
8.   Hunter D, Mati JK. Contraception, Family Planning, and HIV. Paper prepared for the Conference on "AIDS and Reproductive Health", Bellagio, Italy, October 29-November 2, 1990.
9.   Jain AK. Fertility Reduction and the Quality of Family Planning Services. *Studies in Family Planning* 20(1):1-16. 1989.
10. Keller A. Patient Attrition in Five Mexico City Family Planning Clinics, in Stycos J editor, *Clinics, Contraceptives, and Communication.* Des Moines: Meredith Corporation. pp.25-50. 1973.
11. Kreager P. *Family Planning Dropouts Reconsidered: A Critical Review of Research and Research   Findings.* London: International Planned Parenthood Federation. 1977.
12. Liskin L, Quillin W. Long Acting Progestins-Promise and Prospects. *Population Reports.* Series K: Number 2. 1983.
13. Marshall J. Acceptability of Fertility Regulating Methods: Designing Technology to Fit People. *Preventive Medicine* 6:65-73. 1977.
14. Mernissi F. Obstacles to Family Planning Practice in Urban Morocco. *Studies in Family Planning* 6(12):418-425. 1975.
15. Mundigo A. Honduras Revisited: The Clinic and its Clientele. *Clinics, Contraception, and Communication: Evaluation Studies of Family Planning Programs in Four Latin American Countries,* Stycos J, editor. New York: Appleton-Century-Crofts, 1973.
16. Ngin C-S. Reproductive Decisions and Contraceptive Use in a Chinese New Village in Malaysia. Unpublished Ph.D thesis, University of California, Davis, Department of Anthropology, 1985.
17. Pariani Hermiyanto S. Does Choice Make a Difference to Contraceptive Use? Evidence from East Java. *Studies in Family Planning* 22(6):384-390. 1991.
18. Phillips J, *et. al.* Worker-Client Exchanges and the Dynamics of Contraceptive Use in Rural Bangladesh. Paper prepared for 1986 Meeting of the Population Association of America, San Francisco, California.
19. Population Crisis Committee: Access to Birth Control: A World Assessment. *Population* Briefing *Paper*, Number 19. 1987.
20. Rinehart W, Winter J. Injectable Progestogens: Officials Debate but Use Increases. *Population Reports.* Series K: Number 1. Population Information Program of Johns Hopkins University, Baltimore. pp. 1-16. 1975.

21.   Schrimshaw S. Women's Modesty: One Barrier to the Use of Family Planning Clinics in Ecuador, in Marshall J, Polgar S (eds.): *Culture, Natality, and Family Planning*, Chapel Hill: Carolina Population Center, University of North Carolina. pp.167-187.

22.   Shain RN. Acceptability of Contraceptive Methods and Services: a cross cultural perspective, in Pauerstein C, Shain R (eds.): *Fertility Control: Biologic and Behavioral Aspects* Hagerstown: Harper and Row. 1980.

23.   Simmons R, *et. al.* Beyond Supply: The Importance of Female Family Planning Workers in Rural Bangladesh. *Studies in Family Planning* 19(1):29-38.

24.   Sivin I. International Experience with Norplant[R] and Norplant[R]-2 Contraceptives. *Studies in Family Planning* 19(2):81-94. 1988.

25.   United Nations: World Contraceptive Use Chart. 1987.

26.   Wasserheit J. Reproductive Tract Infections in a Family Planning Population in Rural Bangladesh. *Studies in Family Planning* 20(2):69-80. 1989.

27.   Whelan E. Compliance with Contraceptive Regimens. *Studies in Family Planning* 5(11):349-355. 1974.

28.   World Health Organization: *Norplant[R] Contraceptive Subdermal Implants: Managerial and Technical Guidelines.* WHO/MCH/89.17. 1990.

29.   Zimmerman M, *et. al.* Assessing the Acceptability of Norplant[R] in Four Countries: Findings from Focus Group Research. *Studies in Family Planning* 21(2):92-103. 1990.

BRINGING A CLIENT AND LIFE-CYCLE PERSPECTIVE TO SCIENTIFIC EVIDENCE

Commentary on J. Kaufman

Judith Bruce[1]

An appropriate programmatic response to what we have learned in this symposium about the variability of pharmacokinetics of steroidal contraceptives includes the need to revise two key concepts which frame approaches to giving services: "risk" and "effectiveness". In family planning and related reproductive health programs, the scientific and clinical meaning of those concepts needs to be revised to align them more closely with the perceptions of clients who seek and use contraception on a voluntary basis.

My proposal for the revised risk concept is as follows: In voluntary family planning, after medical exclusions are applied, there is one overriding decision maker about risk and that is the client. The client's bases of risk assessment are all valid, though likely expressed in simple ways -- "I don't like the way it feels." If we seek to change a client's risk assessment, we must change the way she feels about a method.

Regarding effectiveness, it has been widely recognized for some time that there are often vast discrepancies between theoretical effectiveness and use effectiveness. Despite the understanding that use effectiveness is the one that counts for the client, many providers still rely on theoretical effectiveness as their guide, and this is often interpreted through their own biases. In presenting effectiveness to clients, it is important that effectiveness be seen as responding to the degree to which a client can comply with the demands of a method.

Despite increasingly high information levels about contraceptives, we find low competence of information among providers and users alike. It is important to remember

---

J. BRUCE • The Population Council, One Dag Hammarskjold Plaza, New York, NY 10017

that there are well documented instances in which the use effectiveness for oral contraceptives falls below that of barrier methods. For example, in the Philippines in a twelve-month period in the mid-1970s the pregnancy rate for oral contraceptive users exceeded that of those who were using periodic abstinence.

If these revised approaches to risk and effectiveness are accepted, they have implications for different levels of activity in our communities of scientists, policy makers and program managers, and direct providers.

1. SCIENTISTS

1)   Client defined "risks" of contraceptives provide an incomplete definition for scientists. We must continue to document and deal with the silent dangers of steroidal contraception -- this is a special challenge as clients themselves don't often experience the most medically serious side effects. And the second challenge is to continue to refine steroidal contraception to respond to perceived but "harmless" side effects as users report them.

2)   The latent (and sometimes overt) demographic rationale for the development of contraceptives and related emphasis on "theoretical" effectiveness has hindered a pluralism in approach. A concept of theoretical effectiveness has fed the notion that one could develop the "perfect method." This symposium affirms tremendous inter- and intra-population variation in the response to steroidal contraceptives. The implication of this pluralism of response is to accept that each available method may serve a smaller share, though potentially more people (as contraceptive use rises), as clients find a method that is optimum for them at a given juncture in their reproductive lives.

Lacking choice and experiencing an unacceptable level of side effects, a number of clients will discontinue use or misuse a method and perhaps as a result, experience either an unwanted birth or an unsafe abortion.

It is important, in the light of the findings of this symposium, to define appropriate goals of success for new methods or for new formulations of old methods. Perceived or real donor demands for a method to have a very large "market share" are shortsighted. Given 30 years of a reproductive span over the course of which a woman may be sexually active and hold three different reproductive intentions -- delaying, spacing, or limiting -- one would assume that we need a variety of methods. Though one method, the oral contraceptive, could theoretically fulfill all these needs, those who are delaying need to delay without inhibiting future fertility. Spacers may wish to combine breast feeding with contraception, and limiters may seek contraception that does not require day-to-day decision making.

In reality, of course, we need far more than three methods because one approach, even if optimized, will not meet everyone's needs. Societies with high contraceptive prevalence are characterized by large amounts of switching of methods. This is true in the United States, in which women typically use three methods over the course of their reproductive lifetimes. In Bangladesh, the Matlab I and II experiments illustrate the important role of switching. In Matlab II, which employed more methods and was explicitly client oriented, not only did levels of acceptance reach higher levels than in Matlab I, but, much more importantly, sustained use was much higher. After 18 months, about two-thirds of the women continuing to use contraception in Matlab II had switched methods.

It may be a paradox of the availability of more and safer methods that each method will have a relatively smaller share cross-sectionally of overall use as clients find the best method for the moment. The development of a new contraceptive should not be premised on the claim that it is going to be right for everyone or most everyone, but rather because it could be a marked improvement in effectiveness or in the health or side-effect profile clients experience, or that it will reach a poorly served group. (Examples include methods to be used by breast feeding women, methods compatible with adolescents' sexual patterns, and the medical abortifacients.)

3)   In the process of contraceptive development, whether steroidal or non-steroidal, we need to promote broader and earlier consultation with prospective users. Women's leadership groups and others may provide a channel for clients' perspectives, but even those who are professionally positioned to link clients with those in the contraceptive-development community get out of touch. It is almost impossible to second guess women's collective or individual points of view. We must consult prospective clients.

As leads are considered in the early clinical phases, and certainly in the introductory phases, clear mechanisms for the feedback from users is essential. We can be more creative at each of these stages. For example, in early clinical trials, a sub-group of subjects could be debriefed in focus groups and/or open-ended questions which would parallel and cross-validate the information derived using more conventional approaches from another subset of clients. This more open-ended approach to research is more likely to yield advanced clues to some of the silent, latent, hard-to-describe side effects -- some potentially attributable to pharmacokinetics -- that become strong features in a client's response in a program setting.

2. POLICY MAKERS

There is much to say here, but let me make two central points:

1)   Our language must be respectful of women when we speak about risk and effectiveness. Earlier in this symposium, people have spoken about women's perceptions of methods -- sometimes with the implication that "they have no right to feel this way" -- and to counter risk fears with the specter of masses of unwanted pregnancies. Instead of being sympathetic to the dilemma women may find themselves in -- tolerating unwanted aspects of contraceptives or facing an unwanted pregnancy -- this approach affronts them. There was an era in which the risks of fertility control, and specifically of oral contraceptives, were trivialized by posing ludicrous comparisons, such as the comparative risks of dying in sporting accidents in the Alps (100 people per year), fatal canoeing accidents in the Shenandoah Valley in a three-month period (17), and the risks of death in one hour of flying a sail plain (4.4 per 100,000 hours). A revised view of risk affirms and seeks to respond to women's perceptions.

If we discount women's responses to rumors, feelings about methods and so forth, then a voluntary system of fertility control will make little progress. A place to begin to resolve the value conflict that often plagues the client/provider relationship is at the international level. There is a principle among those who train that the way a trainer treats the trainees is the way the trainees will treat the people they work with. Hence, if we do not give clients' perspectives a proper weight, how can we expect providers to do so?

2)   A second major task, I believe, for the international community is to acknowledge intra-population and inter-population variation in response to steroidal contraceptives in research agendas. This might include providing substantial amounts of support for non-steroidal contraceptives, and revised procurement and financing policies. As we have discussed in an earlier part of this symposium, there are a number of negative consequences to procurement policies which make price of product the only criterion. There are instances in which donors of contraceptive commodities have switched brands of oral contraceptives over a short period of time and in the process -- albeit unintentionally -- have done tremendous harm to client/provider confidence. If policy allows for the development of a plurality of methods, the financing and procurement policies must do the same, supporting both variety and continuity in supplies.

3. PROGRAM MANAGERS

1)   In many countries, including some of those with the highest levels of unwanted fertility and strongly expressed national concern about population growth, a clear mission statement to guide providers is lacking. Often a country defines the responsibility for solving the "population problem," centering the governmental response in one ministry or sub-ministry health-giving service. It is inappropriate, confusing, and unfair to demand

that a health service provider be responsible for the demographic state of the nation.

It would be of great assistance to providers, and provide a response to the ethical and medical demands of variation among people, to clarify for providers their role as health care givers in a voluntary system. Targets, incentives and so forth militate against an individualized approach to the provision of contraceptives. Paradoxically, it is this individualized approach -- often seen as a "luxury" by governments -- which is the most demographically effective in voluntary systems.

2)   A second suggestion to national level policy makers is that they pursue "epidemiologically sound procurement." Some others have touched on this, so, in brief, it would be very useful to put together a health profile of the people -- particularly women --to be served, their prevailing habits, acceptability studies and so forth when deciding which contraceptives to make available and in what quantities. Many years ago, in my own work in Egypt, I tried to do this and had great difficulty in reconciling an anemia rate of 60% and very high liver disease levels, with the choices the country had made about the contraceptives to be made available. Much has been refined since then, both in health information and in our knowledge of contraceptives. Let's bring that information together.

The goal of such epidemiologically sound procurement policies would be to (a) exclude or minimize the possibility of wrongful prescription and (b) maximize the match between the presumptive health status of the population and the methods available. Examples of excluding wrongful prescription are aggressive efforts by countries to curtail dumping of contraceptives. Or in instances where coercion is an issue at the local level, think carefully about the types of methods made available and whether they are more subject than others to abuse. If it is clear that a high proportion of the population in an area has reproductive tract infections or sexually transmitted diseases that are difficult to detect in normal clinical examination, possibly the IUD is not the best choice, or IUDs should be provided only where STD diagnosis and treatment is reasonably competent.

Another action countries can take in the spirit of epidemiologically sound procurement is to use the media more extensively. We have heard much about how much individuals and couples have learned from the media over the last 20 years about methods, much of it erroneous and fear-provoking. The media can also be used not to promote given contraceptives, but rather to promote the concept of choice -- which would be of value to both providers and clients -- and to provide basic information on methods and outlets. A nation with a high level of media penetration, for instance by way of radio, could think of the media as part of its health-giving system, and indeed part of their safety system as clients are encouraged to refer themselves for care when they show certain symptoms, have their implants or IUDs removed when the term of effectiveness expires, and so forth.

4. PROVIDERS

Joan Kaufman has amply covered this subject. I can add only a decision tree exercise. This is the articulation of the programmatically sound risk and effectiveness concepts at the point of care giving.

The provider has the responsibility to set in motion two kinds of screening processes: (a) the exclusion process and (b) the monitoring process. The exclusion process is that which screens out the risk that can be identified *a priori*. This process should be a dialogue in which what the provider knows and what the client knows are considered in tandem. For example, the provider knows what the clinical exclusion factors are, but the client knows personal and family history. The provider knows relative contra-indications (and in this someday diet may be included should the metabolism of steroids be better understood), and the client knows personal health status and practice. The provider should know, but many do not, the behavioral requirements of the method and possibly some information from acceptability studies which may be useful to the client, while the client knows her daily patterns and the demands of her sexual relationship(s).

This dialogue results in the designation of a method as either being potentially safe and the client being one who could potentially successfully use the method, or being unsafe or unusable and therefore excluded. That is the first major branch.

That is just the beginning. Having excluded the potentially safe from the potentially unsafe, we still are some distance from the identification and support of successful users. Those in the potentially safe category divide into those who accept to try the safest methods for them (we hope that in most systems this is a plural, that there is not simply one), and those who do not accept, for whom we must provide alternatives.

Of those who accept to try the "safest methods," we have two streams of use. In one stream, that which leads to success, all three of the following conditions have to be present; if any one is absent, there will be unsuccessful use. This use takes place across a dimension of time, so the notion of a first single visit in which screening is done and a firm choice is made is clearly a wrong model. The median term use of steroidal contraceptives increases when the client (a) complies (b) can tolerate the side effects and (c) can manage the partner's response.

Another stream of users over time discontinue or fail to prevent pregnancy because (a) they can't comply, (b) they can't tolerate the side effects, or (c) the partner interferes. When any of these things occurs, the care-giving system should be set up so that the client returns to some part of the delivery system to learn more or to change methods.

With regard to this issue of partner's interference, I would like to call your attention to some of what we have learned about the scope of acceptability we found in different parts of the world, most recently through six qualitative studies done in East Africa. In increasing numbers of cases, we are hearing that women are willing to tolerate side effects even if they don't like them because of their high motivation to avoid pregnancy, but that the existence of sickness allows the partners to either (a) detect surreptitious use, (b) change sexual patterns in a way that bothers women or changes fundamentals in the relationship, or (c) insist on a switch. One major issue for the coming years will be to more clearly define men's roles in contraception. Efforts must address the lack of male use of contraceptives and the lack of male support for the female's use of contraception. This, by itself, may be a suitable subject for media attention in a country program.

In summary, and in thinking about the programmatic decision tree listed above, let us review what we know. Do we know:

A. for whom steroidal contraceptives are dangerous?

B. who among those who are "safe" will experience side effects?

C. which side effects they will experience?

D. when in the use segment they will experience side effects; that is, when the onset of symptoms will occur?

E. the intensity of side effects?

F. the duration of side effects? which ones the clients can tolerate?

G. which the partner can tolerate?

With prevailing levels of information, we can include in the exclusion process of care-giving only item A, regarding excluding those for whom steroidal contraceptives are clearly medically contraindicated. In answer to items B-F, these will remain part of a client-managed monitoring process unless and until research into pharmacokinetics and other aspects of steroidal contraception yields more information. F and G may be forever in the hands of the clients. As we learn more, more of these items can be discussed with clients in the initial exclusion process and fewer of them will be left for clients to learn about through experience.

COMMENTARY ON PRESENTATION OF DR. PRAMILLA SENANAYAKE

Gary S. Grubb[1]

Dr. Senanayake has pinpointed many exceptions to the assumptions that occur in clinician-patient interactions and she has shown how these exceptions sometimes should alter the traditional clinical indications and contraindications for a contraceptive. Her review is so comprehensive of the provider's perspective on guiding the choice of a contraceptive, that I plan to offer a commentary that is complementary to her review by giving a contraceptive developer's perspective on how the attributes of the methods themselves guide a woman's choice of contraceptive.

To compress this discussion on method selection, I have tried to simplify - at a great risk of oversimplifying - the absolutely **key** concerns of a woman when choosing a contraceptive method. She might well ask two questions: "What do I have to do to use the method?" (This question entails all the activities she would need to use it effectively and involves the two main issues of convenience and control.) Then she might simply say, "What will it do to me?" as a shorthand for asking, "How well will it protect me from pregnancy and what side effects will I have?"

CONVENIENCE VERSUS CONTROL

In looking at the first question, "What do I have to do to use the methods?", there is a rough continuum of convenience of use among methods, with sterilization being the most convenient and barrier methods the least convenient (Figure 1). The greater

---

G. GRUBB • Family Health International, P.O. Box 13950, Research Triangle Park, NC 27709

convenience among the long-acting steroids of implants and injectables mainly involves not having to remember to do anything other than to return for clinic visits.

Moving down the convenience continuum, problems with inconvenience resulting in incorrect use and user failure begin with the pill. The inconvenience factor has not often been the reason for avoiding initial selection of pills but for discontinuing them after a pill-failure. At the bottom of the continuum, inconvenience has been one of the main factors plaguing the successful use of the coitally-associated methods. There remains the important question of whether there will be any appreciable compliance problems for such methods as the vaginal rings, transdermal patches and transdermal wearable devices.

As a trade-off with the convenience factor, the issue of who controls the removal of the method is an important personal and programmatic concern. All methods fall into one of three categories of who controls removal of the method: Number one, the clinic or

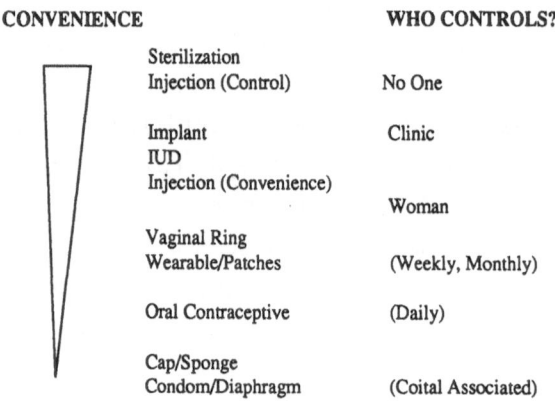

Figure 1. Convenience versus control of contraceptive methods.

the woman. Being able to immediately remove the method is very reassuring to a woman in case she has a side effect - real or perceived - which is intolerable to her. The concern over control of steroids is greater than with other contraceptives because steroids are the only chemical contraceptive meant for systematic absorption. Most women know the chemical permeates the body so it would be logical to them that it could potentially affect any body system and any organ at any time. And we know that, universally, women have great concern about the effects of OCs on their present and future health. So a quick and easy removal of a steroid delivery system would be important to a large segment of women when choosing a contraceptive. In the trade-off between convenience and control, the long-acting steroid methods of the vaginal ring and transdermal devices - and maybe the IUD - seem to present the best compromise to satisfy a woman with concerns about what she has to do and can do to use a method.

EFFECTIVENESS

Now turning to the second question, "What will it do to me?", there is a definite dichotomy in the theoretical effectiveness of methods, with steroids and IUDs being very effective as opposed to all the other methods. Pills are sometimes the exception to the high effectiveness of steroids because of user failure. The degree of noncompliance with correct pill use which could lead to 'pill failure' is only now being discovered by studies using a pill bottle with a computer chip in the lid which records the time it was opened. Initial studies with the device revealed much poorer method compliance than did the pill-taking compliance records the patients kept. In the future, these devices could help women select a method: a clinician could have a new pill user take pills from the device for several months. If the pill-taking data showed the patient did not correctly take the pill and had been at risk of pregnancy, the clinician could recommend another method before a real "pill failure" occurred.

SAFETY AND SIDE EFFECTS

When a woman asks what the contraceptive method will do to her, she is certainly asking about side effects. In the last 15 years, family planning providers have come a long way in avoiding serious side effects by better identifying patients at increased risk of IUD and steroid complications. When a clinician guides a woman in selecting a contraceptive that would minimize side effects, some methods (e.g. IUDs and diaphragms) are fit to the individual's anatomy to minimize the side effects.

However, we know - and this conference is all about how - physiologic differences between individuals are at least as great as their anatomic differences. The clinician typically prescribes OCs in a constricted range of doses of combination OCs intended to prevent almost all ovulations and provide a *very effective* level of protection for all women, despite large physiologic differences in OC metabolism between women.

Because of our lack of knowledge of the relationship between OC dose, resulting serum steroid levels and pharmacodynamic effects, some women will obviously receive much higher drug levels than are needed to be contraceptively effective. Do these higher steroid exposures produce the side effects often seen with the pill? We do not have good evidence for this although this is a widespread clinical assumption among clinicians who adjust pill dosage to alleviate side effects. To show how this differs from a parallel clinical situation, patients experiencing side-effects from drug treatment for asymptomatic disease (e.g., hypertension, hyperlipidemia, glaucoma) have some physiologic measure of whether the drug is working. If the drug is working, the dose is titrated down to minimize the side effects. However, with OC use, the physiologic measures of effectiveness - assessment of ovulation, endometrium and cervical mucus thickening - are not practical in the clinic.

PREDICTING IRREGULAR BLEEDING

A woman may ask about the side effects of steroids because she has never used one. Before committing her to use of a long-acting steroid, wouldn't it be wise to see how well a woman tolerates an oral contraceptive? Are there any data to show that there is a predictive correlation between side effects with a progestin pill or injectable and longer-acting steroids? Even if the data were not productive on a population basis, a woman who experiences one or two months of irregular bleeding with a progestin-only pill or injection would have a much better idea of her tolerance of Norplant^R side-effects, for example, than she would after very comprehensive, careful pre-insertion counseling.

How could a clinician help a woman using a long-acting steroid when she experiences the most common side-effect, irregular bleeding - the one responsible for most method-related discontinuation? With Norplant^R, we simply say to all women that the bleeding should improve after 6 to 9 months. However, we need to carefully analyze the bleeding data in the first few months to try to predict which bleeding patterns are *likely* to improve and which are *unlikely* to improve and fit the "normal" pattern by 6 to 9 months. We have suggestions that there may be other predictors of future bleeding patterns since several investigators treating Norplant^R users are convinced that lower- weight women have more irregular bleeding problems and there is some data to support this. If clinicians could have reasonable predictors that the irregular bleeding would continue, we should run studies to find if modifying the dosage would alleviate the bleeding.

MULTIPLE DOSES OF LONG-ACTING STEROIDS

A reduction of the dosage to avoid discontinuation due to side effects would be made at the expense of a reduction in effectiveness of a method. But why can't a woman decide to have a slightly higher risk of failure with a long-acting steroid contraceptive? It is likely the failure rate would still be less than for OC use failure - which can be 10 - 15%.

Since Norplant^R is more effective than sterilization for women weighing less than 50 kg, the dosage of Norplant^R might be adjusted for those lower-weight Norplant^R users with bleeding irregularities or other side effects. There is a lot of attention given to reduced Norplant^R effectiveness in women more than 70 kg. But, since most Norplant^R users are in Asia, I estimate that there are at least 10 times as many Norplant^R users less than 50 kg than there are more than 70 kilograms. From a public health perspective, side effects in lower-weight Norplant^R users should deserve equal attention.

Inter-individual differences in physiology may overwhelm regional differences in true physiological responses to steroids but regional differences in the **tolerance of** steroid side effects could also result in adjustment of the contraceptive. Rigid standardization of reporting side effects are needed to compare rates of side effects in different centers, countries and regions. But regions which have high discontinuation rates for irregular bleeding, for example, may be appropriate for a dose that optimizes the bleeding patterns without sacrificing much effectiveness.

Why haven't we seen adjustments for individuals in the dose of the long-acting injectables that have been in use for more than 20 years? Besides marketing reasons, one scientific reason may be that DMPA and NET-EN have serum levels that drop off rapidly after injection. So a small reduction in dose would probably provide too little protection at the end of their durations of action. Steroid delivery systems with more even release patterns should be capable of producing systems with at least two release rates and, therefore, the possibility of an adjustment to individuals.

Could multiple doses be made practical for programmatic use? Clinical guidelines would have to be kept simple for deciding which of two doses to start with or change to if there are significant side effects. How feasible would it be to make multiple doses for long-acting steroid contraceptives? A range of doses could be easily produced for an injectable by varying the amount injected. A variable number of multi-capsule implants could be inserted. Areas of transdermal delivery surface on a patch or wearable device could be varied. However, single or double implants, steroid-containing IUDs and vaginal rings would have to be fundamentally modified, so they may be the most difficult to adapt for variable dosing.

In any drug development program, multiple doses are tested. When bringing a product to market, pharmaceutical developers give first priority to getting approval for one dose first. Later, other doses may be submitted for approval. However, I would like to see much more attention to multi-dose development with long-acting steroid contraceptives. Long-acting steroid contraceptives have already greatly reduced the levels of steroid to which women are exposed from the levels found with OC use. I think we should strive to provide multiple dosages of long-acting steroids that will minimize steroid exposure to the *individual* woman as much as possible. Minimizing steroid exposure is the responsible course of development in view of the possibility of avoiding side effects and the serious questions about the effects of long-term steroid use that will continue to be investigated for many years. When there are multiple doses of long-acting steroid contraceptives to choose from, then a clinician will gain a much improved ability to guide a woman's choice of contraceptive.

A CLINICIAN'S PERSPECTIVE

Soledad Díaz[1]

I am a clinician, and I have been following this meeting from that perspective. I confess, however, that I feel like a client who has come for the first time to a family planning clinic. After having listened to the explanation of all the contraceptive methods available, their characteristics, effectiveness, mechanism of action, advantages and disadvantages, etc., I have been asked to make comments and take a decision. As a clinician, how do I transfer what I heard here into practical messages that will be of use to women who come to the clinic and ask "Which contraceptive should I use and which better suits my, let's say, metabolic approach to life?"

We have listed a great many things that can influence the way in which women respond to contraceptive methods or steroid administration. Let me briefly summarize what I have found most provocative.

Diet seems very important. This last paper (see Longcope in this volume) clearly shows that there is an acute affect of diet on estrogen metabolism, and there is also a long-term effect. We should pay attention to both when we record information or when we suggest changes in diet. We tend to make suggestions about diets to clients based on their most recent dietary habits and not regarding a woman's long-term dietary habits. One of the things I learned here is that both of these components of the diet, the long-term habits and

S. DÍAZ • Instituto Chileno de Medicina Reproductiva, Santiago, Chile

the current intake, might affect the metabolic response to contraceptive steroids.   The metabolic pathways of the steroids are altered by diet, because diet affects liver function and the function of the gastrointestinal flora and hence, enterohepatic recirculation.   But the metabolic pathways also have a genetic dimension, and I missed hearing more about this genetic component.   Some populations are rather homogeneous or tend to have a common genetic pattern, and maybe the messages that come from such groups are not valid for other populations.   However, we tend to overlook interpopulation differences and to extrapolate findings from one setting to another.

We have also heard a lot about weight, lean and fat mass, and body composition, and how such variables may influence the outcome of the contraceptive methods.   Since weight is a by-product of a long-term diet, here again diet seems to be very important.   But it is not only weight: it was explained this morning that women may be shaped as "apples" or "pears," and that such patterns of weight distribution have metabolic consequences.   That caught my attention because we have been trying to look at and to explain differences in contraceptive response between different populations and differences in the length of lactational amenorrhea that are not related to the breastfeeding pattern.   I have never thought that a photograph of the women involved in the studies might have helped us to better understand what was happening.

And finally there is the role of sex hormone binding globulin (SHBG).   We have mentioned it several times during this meeting, and we know that SHBG is increased by estrogen levels, and if estrogen levels are affected by diet and SHBG may be affected also directly by diet, there is another component of the equation that requires further research.

How does the route of administration interact with these factors?   Does the gastrointestinal component have importance for drugs that are administered parenterally? What bearing does it have if the oral administration throws a large amount of the drug through the gastrointestinal tract to the liver or if the drug passes through the liver already diluted in the circulation from a depot system?

There is also the problem of endogenous steroid interaction.   When we are using a progestin-only contraceptive, we should remember the drug is interacting with endogenous estrogens.   And the endogenous estrogen levels are not only affected by the amount or the potency of the progestin administered, but also it seems by diet, body composition, liver function and maybe other components of daily life.   I think the best way to organize this array of questions and considerations is by discussing two different practical situations.

NORPLANT[R] subdermal implants

The first is what happens with NORPLANT[R] subdermal implants.   Ovarian activity in NORPLANT[R] subdermal implants users ranges from ovarian suppression to pregnancy and we still don't know all of the variables that influence what happens to which women at what time.   Some women experience ovarian suppression and relatively low endogenous estrogen production. Some show follicular development. We can find ovarian cysts or normal ovulation, and variable degrees of follicular growth are associated with variable estrogen levels and bleeding irregularities. We do know that women who are of a greater weight ovulate with a higher frequency and tend to show lower plasma levels of the progestin than women of lower weight. This happens when they are using a delivery system that provides a rather constant level of the drug for a long time, levels that should have been similar from one woman to the other. The same sorts of variations in ovarian function are observed with the levonorgestrel vaginal ring, which also delivers a constant dose of the steroid.

We know that weight means something, but is it weight itself which determines the different response? The women's diet lead them to these different weights. It may be that the diet that is responsible for the weight is at the same time the cause of a different pattern of endogenous estrogen metabolism. In that case, what may be increasing the rate of ovulation in one group is not the higher weight or the distribution of the progestin, but the amount of estrogen that is metabolized through a pathway that generates more active metabolites that then tend to act at the hypothalamic or pituitary level, with a positive feedback.

Well, so I found a place for this diet here, but I acknowledge that I have never before thought of trying to explain bleeding irregularities, or other side-effects of NORPLANT[R] subdermal implants, through variations in diet. Nor have we studied liver function.

Lactational Amenorrhea

The second situation is related to lactational amenorrhea. I'm going to use some data that show the differences in the length of amenorrhea between two populations. A study done by Peter Howie and his group in Scotland (Howie 1981) showed that Scottish women who are fully nursing with suckling frequency above or around six episodes per day and a suckling duration around 100 minutes maintain high prolactin levels and do not ovulate. In that population, as soon as supplementary food is introduced the suckling frequency decreases, and women start showing ovarian activity and ovulation. In this group no one ovulated while fully nursing with a suckling frequency above six episodes per day.

The situation is very different in Chile. Half of the women who are exclusively breast feeding during the first six months postpartum with a frequency of more than 10 episodes per 24 hours experienced the first menstrual cycle within these six months and 10 percent of these women got pregnant in this interval (Diaz et al 1988). This is the shortest duration of lactational amenorrhea and infertility that has been reported. And ovarian function is recovered in half of the women at a time in which the infant receives nothing but breast milk, not even water.

Actually, we are looking in our country at two different groups within the population: A group of women who start cycling early after delivery in spite of fully nursing, and a group of women who will remain amenorrheic for a longer time. The first group is

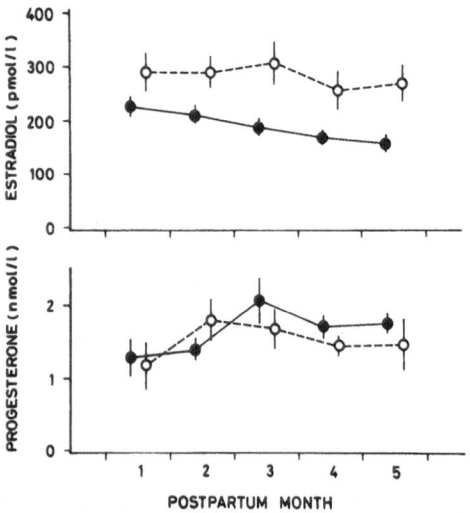

Fig. 1. Plasma E$_2$ and P levels (X ± S.E.) in lactating women who ovulated before (0---0, n = 18) or after (0---0, n = 30) day 180 postpartum. E$_2$ levels were higher in the ovulatory group (p = 0.0001). No differences between groups were detected in P levels. Each dot represents the mean of monthly mean values per subject (Diaz et al 1991).

exposed, if not contracepting, to a high risk of pregnancy during lactation. So it is important to know in advance to which group a woman belongs.

In an effort to identify the differences between these two groups of women, we looked at the early endocrine profile of women who experienced long- or short-term lactational amenorrhea, i.e. early after delivery, when everyone was amenorrheic. The idea was to see if there were early post-partum predictors of the ultimate lactational response. To our surprise, we found early differences in the prolactin response to suckling and the estradiol levels between the two groups of women (Diaz et al 1991).

In the figure above I show the estradiol and progesterone levels (Figure 1) in lactating women. All the samples included here were drawn at least 15 days before the first ovulation, so it excludes the changes related to the first menstrual cycle. Women who ovulated between the fourth and sixth month postpartum had higher early estradiol levels than those who remained amenorrheic for more than 6 months. The women who reinitiate menstrual cycles in the first six months postpartum also had a smaller PRL increase in response to suckling and lower basal PRL levels than the women who remained amenorrheic for a longer period.

The cause of these early differences in prolactin and estradiol levels is not clear. It may be that there are differences in the sensitivity of the breast-hypothalamus-pituitary system to suckling. Is the higher E$_2$ level in the ovulatory group a consequence of this different sensitivity? Let's imagine it is the cause. Where do the different E_2 levels come from? Different peripheral aromatization of androgens due to different composition of body mass? Different absorption of exogenous estrogenic substances? Different clearance of the steroid from the circulation? A different pattern of E_2 metabolism in the liver induced perhaps by differences in the diet or by a genetic component? It is tempting to speculate along this line because Chilean women experience the highest rate of cholestatic disease of pregnancy described in the literature, which involves alterated clearance of steroids by the liver. Is it just a coincidence that they also experience the shortest duration of lactational amenorrhea described? Or is liver function the link between these two conditions?

I am not quite sure about the answers, but I believe that this meeting has invited the liver to be part of the reproductive system.

REFERENCES

Diaz et al. 1989. Lactational amenorrhea and the recovery of ovulation and fertility in fully nursing Chilean women. Contraception 38:53-67.

Diaz, S. et al. 1991. Early difference in the endocrine profile of long and short lactational amenorrhea. Journal of Clinical Endocrinology and Metabolism 72:196-201.

Howie et al. 1981. Effect of supplementary food on suckling patterns and ovarian activity during lactation. British Medical Journal 283:757-763.

REGIONAL POPULATION DIFFERENCES AND POPULATION PHARMACOKINETICS OF STEROIDAL CONTRACEPTIVES

Sang Guo-wei[1]

1. INTRODUCTION

Clinical pharmacokinetics of contraceptive steroids is a challenging discipline with a strong theoretical framework for application to the development of steroidal contraceptives and delivery systems for fertility regulation. Pharmacokinetic parameters of contraceptives are of critical importance in assessing contraceptive efficacy, side-effects and menstrual bleeding patterns which are relevant to acceptability and continuation rates.

However, studies of drug disposition in a number of individuals generally reveal that the essential pharmacokinetic parameters (e.g., bioavailability, volume of distribution, clearance) lie within a restricted range of values. This is especially true if the study group is homogeneous with regard to individual characteristics which may influence the disposition of steroidal contraceptives. Recently, it has been widely accepted that the use of traditional pharmacokinetic modelling and parameters to estimate contraceptive steroids levels in subjects, even with computer assistance, was limited in its accuracy of prediction. This was attributed to large inter- and intraindividual variability in pharmacokinetic parameters.[1,2] The focus of most pharmacokinetic researches has been on evaluating and characterizing interindividual variation. However, significant intraindividual variation has been demonstrated for several contraceptive steroids such as levonorgestrel and ethynylestradiol.[2,3,4] In some cases, intraindividual variability may compare with, or even exceed, the interindividual variability.

G-W. SANG • Family Planning Research Institute, Zhejiang Academy of Medical Sciences, Hangzhou, Zhejiang 310013, People's Republic of China

Steroid Contraceptives and Women's Response,
Edited by R. Snow and P. Hall, Plenum Press, New York, 1994

Realizing the above problems, the previous regional or population comparisons of contraceptive steroid pharmacokinetics appear to have limited significance in explaining variability. However, population pharmacokinetics can entail the summary of pharmacokinetic studies in groups of individuals and the establishment of relationships between individual characteristics and pharmacokinetic parameters. Individual pharmacokinetic parameters quantify the pharmacokinetics of an individual, while population pharmacokinetic parameters quantify the mean population kinetics, interindividual kinetic variability, and residual variability, including intraindividual variability and measurement error.[4] Therefore, it will provide invaluable documentation of population parameters for contraceptive kinetics. Nonlinear Mixed Effects Modelling (NONMEM) is a recently introduced technique of population pharmacokinetic estimation.[5]

This article will discuss the population differences in the pharmacokinetics of long-acting steroidal contraceptives and the need for population pharmacokinetics, from data mainly derived from our laboratory. Then general methodology and the specific data required for population pharmacokinetics will be reviewed.

2. POSSIBLE POPULATION DIFFERENCES IN PHARMACOKINETICS OF LONG-ACTING STEROIDAL CONTRACEPTIVES

In recent years, as multicentred studies develop there are more and more indications that regional and population differences as well as intra- and interindividual variabilities exist in the pharmacokinetics of steroidal contraceptives.

2.1 Pharmacokinetic study on norethisterone enanthate (NET-EN)

The pharmacokinetic profiles of NET-EN and norethisterone (NET) after intramuscular injection of 200mg NET-EN (Schering AG products) were studied in 9 British women and 25 Chinese women by the same investigator using the same methodology.[6] No significant differences in absorption kinetics of NET between British and Chinese women were observed. However, the elimination rate of NET was significantly slower in Chinese women than that in British women. Chinese women showed a greater area under the serum NET concentration - time curve (AUC) and a longer period to decrease serum concentration below 0.335 nmol/1 (**Table 1**).

2.2 Norplant[R]

Serum levonorgestrel (LNG) concentrations after insertion of Norplant[R] were studied in ten healthy, age 26-33 years Chinese women in our laboratory. Their mean body weight and Quetelet Index were 54.7 ± 4.2 kg and 21.0 ± 2.0 kg/m^2, respectively. The in vivo release rates were calculated from recovered implants from other 68 Norplant[R] users.[7] The mean release rate was 304 ug/day during first three months, 95 ug/day from 6 to 12 months, 56 ug/day around 18 months, 51 ug/day at 30 months and 40 ug/day thereafter. These values are significantly greater than that reported by Nash in Chapter 10 of this book.[8] (**Table 2, Figure 1**)

Measurement of serum concentrations of LNG after insertion of Norplant[R] showed that the peak value of LNG was 9.4 nmol/1, which was reached within 3(+3) days in Chinese women. LNG rapidly decreased to 1.9 nmol/1 during the first month, then

Table 1. Pharamacokinetic parameters of NET in Chinese and British women.

Parameters	Chinese (n=25)	British (n=9)	P
Peak value (nmol/l)	42.2 ± 15.4	37.2 ± 24.8	>0.05
Time to reach the peak value	5.4 ± 2.0	4.3 ± 2.2	>0.05
Apparent elimination half-life	14.8 ± 3.8	11.6 ±4.5	<0.05
AUC (nmol/l/d)	725.1	480.8	<0.01
Days for NET levels to decrease <0.335nmol/l	11 ± 21	74 ± 20	<0.01

(mean with SD)

fluctuated around a plateau value of 2.0 nmol/1 for up to 2 years. These blood levels are all consistently higher than that reported by Nash, who found that plasma LNG concentrations normalized to 60 kg body weight were 1.04 nmol/1 at 12 months and 0.98 nmol/1 at 24 months after placement.

Affandi B et al.[9] measured LNG concentrations from 208 Indonesian women who used Norplant[R]. The peak value was approximately 7 nmol/1. LNG levels rapidly decreased during first month, the decrease continued to 10 month later and reached a trough

Table 2. In vivo release rates on Norplant[R] after insertion.

Month	ug/day*	Month	ug/day**
3	75	0-3	304
9	53	6-10	95
18	36	16-20	56
30	29	26-30	51
>30	29	>30	40

*data from Dr. Nash   **from our data

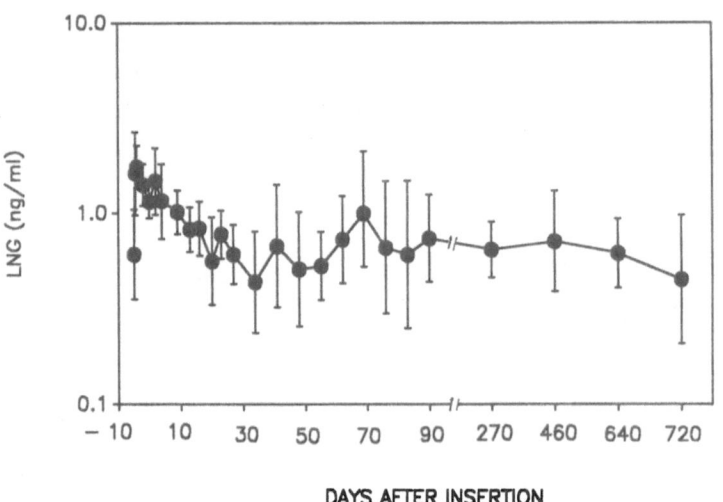

DAYS AFTER INSERTION

Figure 1. Geometric means and 95% C.I. of serum LNG levels after insertion of Norplant[R] in Chinese women.

concentration of 1.1 nmol/1. The LNG concentration then increased again to a plateau of 1.5 nmol/1 around 2 years after insertion.

These data from above three countries clearly showed different pharmacokinetic profiles as well as different plasma LNG levels in women using Norplant[R]. Some differences among populations may be caused by different study design, such as time and frequency of blood sampling, different RIA kits used, and measurements by different laboratories, but not all. Chinese women also showed a higher incidence of irregular bleeding and a lower rate of unexpected pregnancy. Thus there may be a relationship between pharmacokinetic and pharmacodynamic population variability.

2.3 Pharmacokinetics of monthly injectable NET-EN/estradiol valerate

Data from a multicentre pharmacokinetic study on different doses NET-EN with or without estradiol valerate (EV) also showed certain difference between Chinese and British healthy ovulating women.[10] Comparing the data from two centers in Hangzhou and London, peak value of NET (Cmax) were markedly higher in Chinese women for all of four dose preparations (**Table 3**).

In addition, the relationship between dose and serum NET concentration seems to be different in Chinese and British women. It may indicate that Chinese women seemed to be more sensitive to the dose of both progestogen and estrogen.

Table 3. Pharamcokinetic data of NET in Chinese and British.

Dose combination	C_{max} (nmol/l)		$T_{c<0.335nmol/l}$	
	Hangzhou	London	Hangzhou	London
NET-EN 50mg +EV 5mg	15.9 (11.8-21.5)	4.7** 2.1-7.1)	92.2 (79.2-106.9)	68.1* (52.4-88.7)
NET-EN 25mg +EV 2.5mg	7.9 (5.6-11.2)	4.1** (2.0-8.5)	60.6 (47.6-93.1)	67.8 (55.5-82.7)
NET-EN 50mg	12.6	6.5*	91.7	72.4
NET-EN 25mg	8.0 (4.8-13.3)	2.9* (1.8-4.5)	60.8 (43.9-84.2)	57.7 (41.6-71.9)

Mean with 95% C.I.   *$P<0.05$ **$P<0.01$

2.4 Pharmacokinetic profile of DMPA

Population difference in dose - serum concentration relationship was also showed in a multicentre study on DMPA. Analysis of the data from a pharmacokinetic and pharmacodynamic study on DMPA at doses of 25, 50, 100 and 150 mg in Bangkok and Mexico City were conducted[11]. **Figure 1** and **2** showed the Cmax and AUC at different doses of DMPA for both Thai and Mexican women. It is clear that the Cmax and AUC for DMPA in Thai women showed a certain dose relationship, but this was not seen in Mexican women. It was also found that Cmax values were significantly higher in Thai women than those in Mexican women received same dose of DMPA. This kind of population difference in pharmacokinetics of DMPA could not be explained simply by body weight differences (**Figure 2, 3**).

3. FACTORS CONTRIBUTING TO VARIABILITY IN PHARMACOKINETICS OF LONG-ACTING STEROIDAL CONTRACEPTIVES

During the past 25 years, a large number of pharmacokinetic studies on contraceptive steroids have been performed. Almost all of these studies showed large intra- and interindividual differences of pharmacokinetic parameters.[12] Many factors contribute to the pharmacokinetic variability of steroidal contraceptives (**Table 4**).

3.1 Interindividual variability

In the study on pharmacokinetics of NET and NET-EN after injection of 200mg NET-EN carried in 25 Chinese women, the Cmax of NET ranged from 6.5 ng/ml to 25.9 ng/ml, with a coefficient of variation (CV) of 36.8%. The Cmax of NET-EN ranged from 2.2 to 9.0 ng/ml with a CV of 36.2%. The Cmax of both Net and NET-EN showed a four-fold difference. The individual value for the time to reach peak concentration ranged from 2 to 10 days with a CV of 37.0%. The elimination process also displayed wide interindividual differences. Four women showed faster elimination than the others, with serum NET levels decreasing to below 0.2 ng/ml, 56 days after injection. The other 6 women showed slower elimination with serum levels remaining at 0.36 - 1.0 ng/ml, 84 days after injection (**Figure 4**).

These two sub-groups showed clearly different concentration - time profiles. Genetic factors may possibly contribute substantially to the significantly different metabolic clearance of NET since no differences regarding weight, height or Quetelet Index were found between these two sub-groups.

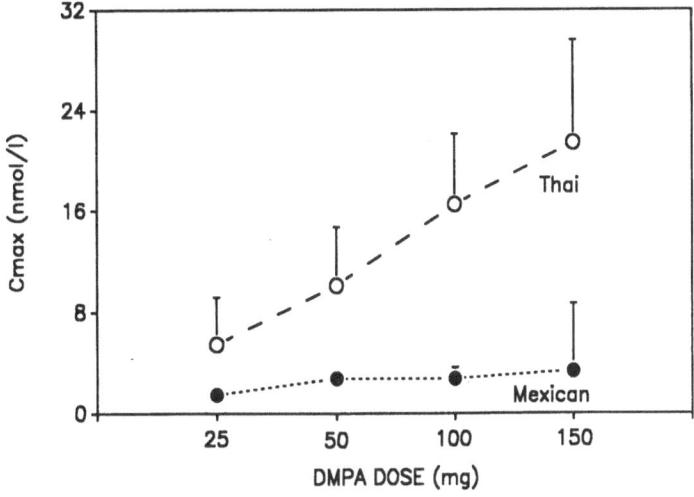

Figure 2. The relationship between C_{max} and dose of DMPA in Thai and Mexican women.

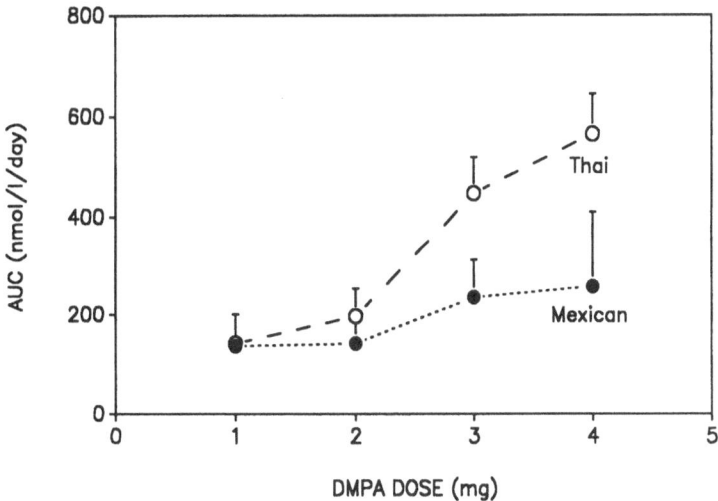

Figure 3. The relationship between AUC and Dose of DMPA in Thai and Mexican women.

Table 4. Factors contributing to variability in steroid contraceptives.

---

*	Body Weight/Size: TBW, ECF, Vd, organ size, function, blood flow, obesity and thinness;
*	Protein levels: SHBG, albumin, ceruloplasmin, α-acid glucoprotein
*	Genetic Factors: Metabolic clearance
*	Drug interaction
*	Disease

3.2 Intraindividual variability

In order to ascertain the pharmacokinetic profile of NET-EN after the long-term use of the monthly injectable NET-EN 50mg/EV 5mg, a longitudinal pharmacokinetic study of NET after the first, sixth and twelfth injections was carried out in 17 healthy Chinese female volunteers. The pharmacokinetic parameters of NET-EN are listed in **Table 5**. There was a tendency towards a decrease in Cmax as the number of injections increased (**Figure 5**). These data showed an intraindividual variation in Cmax of NET from cycle to cycle after multiple injection.

3.3 Body weight

In the mid-1980's, Dr. Odlind of Uppsala University, Sweden, carried out a pharmacokinetic study on two preparations containing 150mg depot-medroxy-progesterone acetate - Depo Provera[R], Upjohn Company, and Gestapuran depot Leo Pharmaceutical products.[13] A significant negative correlation existed between Cmax and body weight ($P \ll 0.05$), AUC and body weight ($P < 0.02$) for both groups (**Figure 6,7**). The results indicated significant differences in the AUC of DMPA between the two groups, the values were 279.4 ± 17.2 nmol/1/d and 221.1 ± 15.0 nmol/1/d, respectively. Investigators then found that 25% of variation in AUC could be ascribed to variations in weight. After correction, the AUC values were 264.8 ± 14.8 nmol/1/d and 237.2 ± 13.3 nmol/1/d, and showed no significant difference between these two groups.

3.4 Sex hormone binding globulin (SHBG)

Several studies have showed that SHBG concentration is related to administration of certain synthetic progestogens and/or estrogens.[14]　On the other hand, many steroidal compounds can bind to SHBG, thereby the pharmacokinetic profile of these steroids will be affected by the SHBG concentration.　A pharmacokinetic study on LNG in healthy fertile Chinese women who received three oral doses of LNG 6mg plus quinestrol (CEE) 3mg at 23 days intervals was performed in our laboratory.　No significant differences of mean elimination half-lives after three administrations were observed.　However, the Cmax and AUC of LNG increased by 5-6 fold after the second administration, and was obviously related to serum SHBG levels (**Table 6, Figure 8**).

4. THE NEED TO UNDERTAKE POPULATION PHARMACOKINETIC STUDIES

From our studies on clinical pharmacokinetics of long-acting steroidal contraceptives given orally and intramuscularly, we can conclude that contraceptive steroids, like all drugs, exhibit very marked pharmacokinetic variability.　With this profound intra- and interindividual variability in pharmacokinetics, questions about the clinical significance of previous regional or population comparisons in contraceptive pharmacokinetics have been raised. In addition, it is important that the estimates of pharmacokinetic parameters are obtained from representative data collected in the population actually treated with the contraceptives since pharmacokinetic parameters may show such large differences between different subject populations.　As we know, individual pharmacokinetic parameters only quantify the pharmacokinetic parameters of an individual, while population pharmacokinetic parameters can　quantify mean population kinetics, interindividual variability, and residual intraindividual variability.　Therefore, population pharmacokinetic studies appear to be essential to document the suggested regional or population differences in the pharmacokinetics of steroidal contraceptives.

4.1 Definitions of population pharmacokinetic parameters

1)　Fixed-effects parameters: these include population-typical values (means) as well as the coefficient of regression relationships between pharmacokinetic parameters and various independent variables (subject characteristics), such as (a) age, weight, height, body composition and sex; (b) underlying pathology and (c) other influences on drug disposition such as concomitant drug therapy, smoking habits and alcohol intake.

2)　Random-effects parameters: these are population variability values (standard deviations) representing interindividual deviation from fixed-effects parameter estimates after the population relationship have been taken into account, and residual random error.

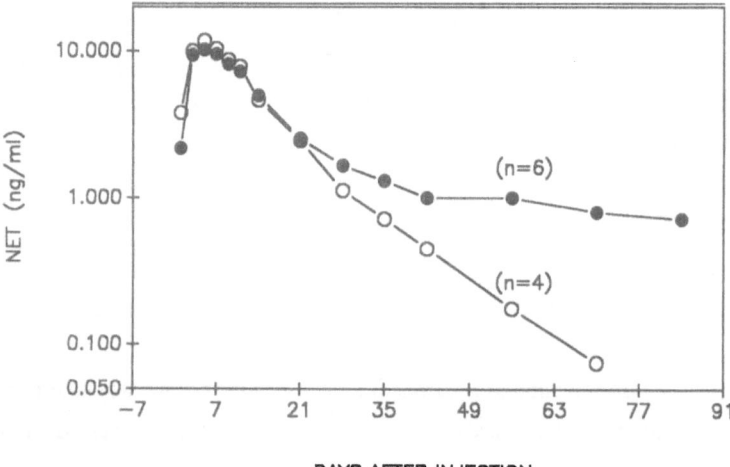

Figure 4. Elimination pharmacokinetics of NET in two sub-groups .

Table 5. Pharamcokinetic parameters of NET-EN after multiple injections of NET-EN 50mg/EV 5mg (mean ± S.D.).

Parameters	Treatment cycle			P
	1	6	12	
Time to reach reach peak value (d)	5.5 ± 1.1	5.1 ± 1.3	5.0 ± 1.4	>0.05
Cmax (ng/ml)	8.62 ± 2.63	6.74 ± 1.91	5.90 ± 1.24	<0.001
AUC (ng/ml/d)	102 ± 29	92 ± 26	86 ± 19	>0.1
Apparent elimin-ation half-life (d)	5.6 ± 1.4	7.0 ± 1.7	6.6 ± 1.7	<0.05
Apparent absorption half-life	1.5 ± 0.4	2.4 ± 1.0	2.3 ± 0.7	<0.05
C at day 28 (ng/ml)	0.64 ± 0.36	0.90 ± 0.39	0.67 ± 0.29	>0.1

3)   Residual error: this includes measurement error and intraindividual variability. The error term expresses the fact that variability in the assay used to measure the drug concentration and moment to moment changes in drug disposition will produce variability in the observed concentration.[15]

4.2 Determination of population pharmacokinetics

Individual pharmacokinetics are estimated by fitting individual data to a pharmacokinetic model.   However, there are several different methods of determining population pharmacokinetics (**Table 7**).   Traditionally, population pharmacokinetic parameters are estimated either by fitting data from all individuals together assuming no

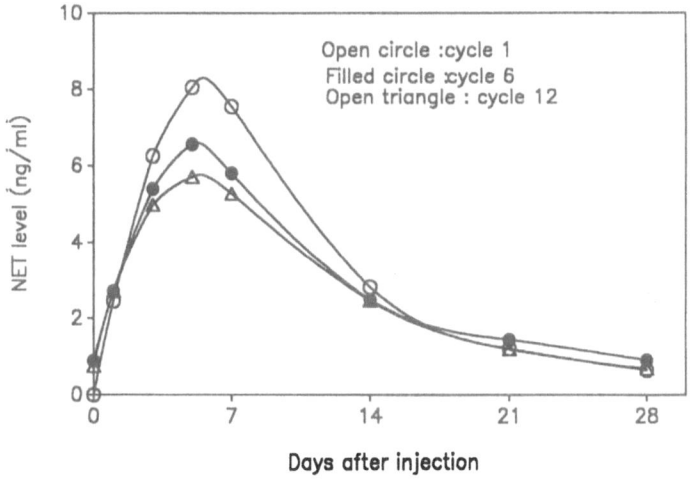

Figure 5. Mean serum NET levels after 1st, 6th and 12th injection of NET-EN 50mg/EV 5mg.

individual kinetic differences (the naive pooled data approach, NPD), or by fitting each individual's data separately, and then combining the individual parameter estimates (the two-stage approach, TS).

The major problem with the naive pooled data approach is that it ignores individuals entirely. All deviations of fit from the model are lumped together in one single error term. Therefore, the method can not estimate the distinct pharmacokinetic parameters corresponding to individual variability in each parameter nor can it estimate the residual variability due to intraindividual variability and measurement error. The method fails to distinguish

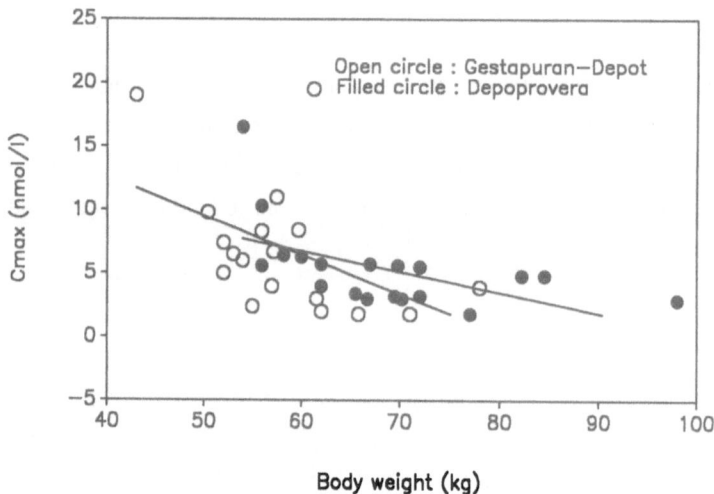

Figure 6. The relationship between C_{max} and body weight of DMPA user (data from Uppsala study).

Table 6. The pharamacokinetic parameters of LNG after 3 doses of LNG 6mg plus CEE 3mg (mean ± S.D.).

Cycle	c_{max} (ng/ml)	Ke (d^{-1})	AUC (ng/ml/d)	$T_{c<0.1ng/ml}$
1	10.98 ±5.12	0.4399 ± 0.1012	40.7 ± 16.5	12.1 ± 3.2
2	61.23 ± 19.30	0.4591 ± 0.1178	217.3 ± 55.2	15.6 ± 3.3
3	67.21 ± 23.50	0.4248 ± 0.0808	246.3 ± 70.0	16.7 ± 3.0

random inter- from intraindividual variability, and it provides no separate estimates of the magnitudes of these effects.[16]

The two-stage method is the opposite of the naive pooled data approach. At the first stage, it regards each individual as completely distinct from all others, and estimates each individual's pharmacokinetic parameters from his data alone. In the second stage, these individual estimates are combined to yield population parameter estimates. But a fundamental problem with the two stage method arises in the way it estimates random interindividual effect parameters. In general, the standard deviation of the individual parameter estimates will overestimate biological parameter variability. This is because each parameter is estimated from the original drug level vs. time data with some error. This error adds variability to the parameter estimates that is not biological in origin. Hence the random interindividual variability will be overestimated.[16]

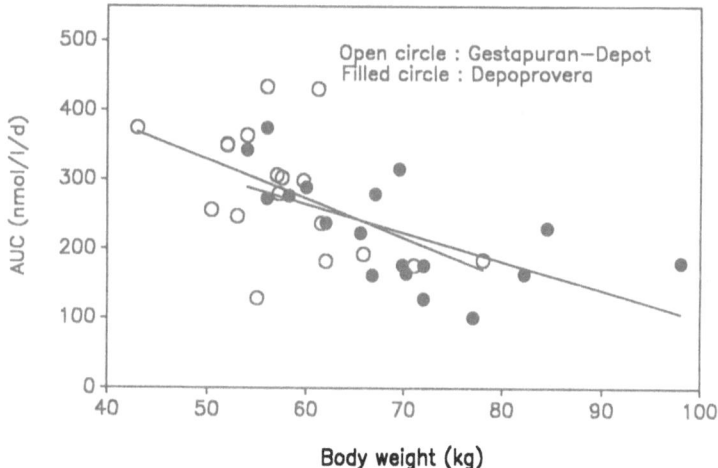

Figure 7. The relationship between AUC and body weight of DMPA user (data from Uppsala study).

To be entirely fair to the two stage method, we believe that the two stage method enjoys several positive features. Least-squares nonlinear regression is a very familiar technique which is understood by many pharmacokineticists. When applied properly, it has proven to be a reliable method of estimating pharmacokinetic parameters in experimental studies. The computer routines for performing the two stage method are available on a variety of computers. When sufficient data are available in each individual to obtain statistically precise estimates of individual pharmacokinetic parameters, and a large number of individuals are included in the analysis, two stage analysis of data from many individuals provide reasonable estimates of population pharmacokinetic parameter distribution. However, the population pharmacokinetic information from the two stage method often comes mainly from studies of healthy volunteers or small numbers   of   subjects   who

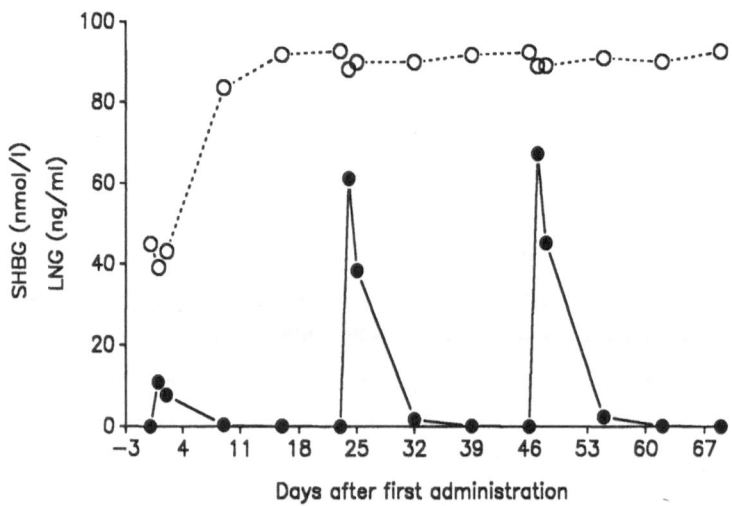

Figure 8. Serum LNG (●) and SHBG (○) levels after receiving multiple doses of LNG 6mg/
CEE 3mg.

Table 7. Determination of population pharmacokinetics.

---

*   Naive pooled data (NPD);

*   Standard Two-Stage (STS) method;

*   Nonlinear Mixed Effects Modelling;

        a. Fixed-effect parameter (population mean, coefficients of regression)
        b. Random-effects parameters (population variability, interindividual
        deviation, residual random error)
        c. Main strength
        scanty data from actual clinical trial can be analyzed

*   NONMEM program

---

inadequately represent those undergoing routine treatment for contraception. Therefore, information generated by the two stage method constitutes a very limited foundation upon which to document any possible regional or population differences in pharmacokinetics of steroidal contraceptives.

Mixed Effects Modelling is a recently introduced technique of population pharmacokinetic estimation which was developed specifically to rectify drawback inherent in the two stage method.[17] Mixed effect modelling allows direct estimation of population pharmacokinetic parameters in a single stage of analysis applied simultaneously to data from many individuals (for details, see methods, below). A main strength of mixed effects modelling is its ability to accommodate scanty data from actual clinical treatment, thus enhancing the use of clinical data for incorporation into techniques of population pharmacokinetics estimation, and clinical pharmacokinetic forecasting for possible population differences in pharmacokinetics/pharmacodynamics of steroidal contraceptives.

5. GENERAL METHODOLOGY FOR ESTIMATING OF POPULATION PHARMACOKINETIC PARAMETERS

Population pharmacokinetic studies is the estimation of population-typical (e.g., mean) and population - variability (e.g., standard deviation) values for each pharmaco-kinetic parameter. The population mean and standard deviation summarize the population distribution of pharmacokinetic parameters. Population pharmacokinetics should be studied in a heterogeneous group of individuals exhibiting a range of values of subject characteristics which are thought to influence drug disposition. This is done deliberately to establish relationships between individual subject characteristics and populations pharmacokinetic parameter distributions.

Two classic methods (NPD and TS) have been thoroughly studied and are not in general, satisfactory. More recently, an alternative approach to population pharmacokinetic data analysis has been implemented in the Nonlinear Mixed Effects Model (NONMEM) computer program.[18] This approach treats the population as the unit of analysis, rather than the individual, and in general requires fewer data points per individual (but many more individuals) than are normally required. By adopting this policy, a much more representative sample of the target population can be obtained and quantitative relationships between pharmacokinetic parameters and pathophysiological features can be investigated in a single step. Until recently NONMEM was mainframe-bound, expensive to run, batch-mode only, and difficult to use. Fortunately, a Fortran-77 version of NONMEM is now available, which can run on appropriately configured minicomputers or microcomputers. Currently, NONMEM appears to provide the most acceptable method.

NONMEM describes the observed concentration-time data in terms of:

1.   A number of fixed effect parameters,θ_k, which may include the mean values of the relevant structural pharmacokinetic model parameters or a number of parameters which relate the structural model parameters to demographic and pathophysiological variables.

2.   Two types of random effect parameters:
a) the variances of the θ_k or the individual structural parameters, i.e., the intersubject variability within the population,$^w\theta^2_k$, and
b) the residual intrasubject variability due to random fluctuations in an individual's parameter values, measurement error and all sources of error not accounted for by the other parameters,σ_ε^2.[19]

The function of NONMEM is to obtain estimates of the average values of the parameters, θ_k, the intersubject variances $^w\theta^2_k$, and the residual (intrasubject) variance σ_ε^2. In practice, steroid concentrations obtained at random (but known) intervals after dosing can be used to model population pharmacokinetic parameters. This method allows the inclusion of steroid measurements made at different times both within and across individuals and also permits varying numbers of sample points per individual. Analysis is based on the premise that population parameters can be modeled as random variables, with estimates of the pharmacokinetic parameters for any individual as a realization of the random variable distribution. An estimate of the deviation of a particular individual from the 'typical' value for a population pharmacokinetic parameter is obtained by the use of a regression equation to model the relationship between individual characteristics and pharmacokinetic parameters. This allows for a decomposition of the variance (or error) in predicting the actual serum levels obtained for an individual as a summation of variance caused by 'fixed effects' (i.e., individual characteristics linked by regression equation), residual interindividual variance, and intraindividual variance (i.e., the error in prediction remaining after accounting for variance between individuals).

The analysis of data using NONMEM requires the development of a pharmaco-statistical model for each of the three types of population parameters to be estimated. This pharmacostatistical model includes a pharmacokinetic model for generating predicted steroid concentration as a function of individual pharmacokinetic parameters, and additional models for the mean and magnitude of variability. The models can be divided into two basic types: structural models, including the pharmacokinetic model and regression formulae for investigating the effect of various fixed effects on pharmacokinetic parameters, and statistical models for variability, including both interindividual variability and residual variability.

6. SPECIFIC DATA REQUIRED FOR POPULATION PHARMACOKINETICS

The concept of collecting data 'routinely' is fundamental to population pharmacokinetics. Routine patient data are essentially costless, abundant, and representative of the patient group of greatest interest: those receiving steroidal contraceptive for contraception. In addition, making sure that the appropriate data are collected is extremely important as this has considerable bearing on the success of population pharmacokinetic studies and determines: (a) what can be learned from existing (retrospective) data; and (b) what prospective data are required to answer specific questions.[19] In a prospective study, specific data needs are determined prior to study initiation and modifications are incorporated into the protocol design and implementation to ensure appropriate data collection. In a retrospective study, population pharmacokinetic analysis is attempted using data assembled after the clinical study has been completed. Typically this involves an attempted analysis without input into study design or data collection.[20]

In general, two types of data, kinetic and demographic are required for population pharmacokinetics.

6.1 Kinetic data

1. Specifying dosage regimen: dose, route, dosage interval, proceeding relevant dosage history
2. Concentration - time data

Four categories of kinetic data

1). Steady-state trough concentration - only yield minimal information
2). Average steady state concentrations
3). Concentrations at any time after an oral dose - several (3 or 4) concentration-time pairs per subject
4). Concentrations at any time after both intravenous and oral dose

6.2 Demographic data

1. At beginning: age, sex, weight, height, smoking, alcohol, co-medication, biochemical/hematological indices
2. During any dosage interval - interesting changes related to treatment

6.3 Data set construction

A NONMEM input data set includes information regarding determined concentrations, times of blood sampling, and dosage regimen including both times and amount received. In order to ensure the accurate recording of this data, collection forms should be designed and included in the evaluation form packet of the prospective study evaluation. In the retrospective study, the dosing regimen data required for construction of the NONMEM data set are based upon the subjects recorded total daily dose and assuming complete compliance to protocol.[20]

6.4 Sample size

No hard and fast rules can be laid down at the moment but common sense dictates that high degrees of interindividual variability can only be explained if a relatively large number of subjects expressing that variability is studied. Moreover, an estimate of the intraindividual component can only be obtained by collecting several (3 or 4) samples per subject from the majority of subjects. A recent study of population pharmacokinetic design indicated that an increase in the number of subjects from 33 to 55 (for three point sampling) can practically eliminate bias and improve the precision.[21]

References

1.   Orme, M.L'E.: The clinical pharmacology of oral contraceptive steroids. *Br. J. Clin. Pharmacology* 14:31, 1982.

2.   Fotherby, K.: Pharmacokinetics of gestagens: Some problems. *Am. J. Obstet. Gynecol.*, 163:323, 1990.

3.   Shi, Y.E., Zheng, S.H., Zhu, Y.H., et al.: Pharmacokinetic study of levonorgestrel used as a postcoital agent. *Contraception* 37:359, 1988.

4.   Brody, S.A., Turkes, A. and Goldzieher, J.W.: Pharmacokinetics of three bioequivalent norethindrone/mestranol 50 ug and three norethindrone/ethynyl estradiol 35 ug formulations: are "low-dose" pills really lower? *Contraception* 40:269, 1989.

5.   NONMEM Project Group, 555 Sciences, University of California, San Francisco, CA 94142.

6.   Sang, G.W., Liu, X.H., Shao, Q.X., et. al.: Pharmacokinetics of norethindrone enanthate 200 mg after intramuscular injection in 25 Chinese women. *Acta Pharmacologica Sinica* 12:184, 1991.

7.   Su, M., Shao, Q.X., et al.: Pharmacokinetics of Norplant[R] in Chinese women. *Beijing Medical Journal* 13:150, 1991.

8.   Nash, H.A. and Robertson, D.N.: Pharmacokinetics of Norplant[R]. Presentation in: Workshop - "Steroid Contraceptives and Women's Response: Regional Variability in Side-effects and Steroid Pharmacokinetics". 21-24 Oct., 1990, Exeter, NH, USA.

9.   Affandi, B., Cekan, S.Z., Boonkasemsanti, W., et. al.: The interaction between sex hormone binding globulin and levenorgestrel released from Norplant[R], an implantable contraceptive. *Contraception* 35:135, 1987.

10.   World Health Organization, Task Force on Long-acting Systemic Agents for Fertility Regulation, Special Programme of Research, Development Training in Human Reproduction: A multicentred pharmacokinetic/pharmacodynamic study of monthly doses of norethisterone oenanthate with or without oestradiol valerate. (To be published).

11.   Sang, G.W. and Garza-Flores: A review of pharmacokinetic data generated on Depo-provera in different populations. Presentation to: Steering Committee Meeting of Task Force on Long-acting Systemic Agents for Fertility Regulation, Special Programme

of Research, Development and Research Training in Human Reproduction, World Health Organization, Geneva, April, 1987.

12.   Fotherby, K.: A critical evaluation of the pharmacokinetics of contraceptive steroids. Presentation in: Workshop - "Steroid Contraceptives and Women's Response: Regional Variability in Side-effects and Steroid Pharmacokinetics". 21-24 Oct., 1990, Exeter, NH, USA.

13.   Odlind, V., et al.: Bioavailability of two depot medroxyprogestrone acetate preparations - Gestapuran depot and Depo-Provera. Personal communication, 1987.

14.   Orme, M.L'E., Back, D.J. and Brekenridge, M.: Clinical pharmacokinetics of oral contraceptive steroids. *Clinical Pharmacokinetics* 8:95, 1983.

15.   Sheiner, L.B. and Grasela, T.H.: An introduction to mixed effect modelling. Presentation in: Symposium - "The Application of Population Pharmacokinetics to Drug Development and Utilization", Nov. 4-5, 1986, Washington, DC, USA.

16.   Sheiner, L.B. and Beal, S.L.: Evaluation of methods for estimating population pharmacokinetic parameters. II. Biexponential model and experimental pharmacokinetic data. *J. Pharmacokinetics and Biopharmaceutics* 9:635, 1981.

17.   Sheiner, L.B., Rosenberg, B. and Marathe, V.V.: Estimation of population characteristics of pharmacokinetic parameters from routine clinical data. *J. Pharmacokinetics and Biopharmaceutics* 5:445, 1977.

18.   Beal, S.L. and Sheiner, L.B.: NONMEM users guide. San Francisco: University of California, 1980.

19.   Whiting, B., Kelman, A.W., Grevel, J.: Population pharmacokinetics: Theory and clinical application. *Clinical Pharmacokinetics* 11:387, 1986.

20.   Antal, E.J., Grasela, T.H. and Smith, R.B.: The application of population pharmacokinetic analysis to large scale clinical efficacy trials. Presentation in: Symposium - "The Application of Population Pharmacokinetics to Drug Development and Utilization", Nov. 4-5, 1986, Washington, DC, USA.

21.   Al-Banna, M.K., Kelman, A.W. and Whiting, B.: Experimental design and efficient parameter estimation in population pharmacokinetics. *J. Pharmacokinetics and Biopharmaceutics* 18:361, 1990.

VARIABILITY IN STEROID PHARMACOKINETICS: NEEDS FOR FUTURE RESEARCH

Laneta J. Dorflinger[1]

1. INTRODUCTION

Research presented during this workshop documents the variability of estrogen metabolism among women, particularly as it relates to diet and body fat topography (Bradlow, Longcope, Goldin in this volume), and the large degree of variability in the pharmacokinetics of contraceptive steroids (Goldzeiher, Fotherby, Garza-Flores in this volume). These latter presentations on synthetic estrogen and progestin pharmacokinetics described intraindividual and interindividual variation, both within a given ethnic population and across different ethnic populations. The potential link between the variability in endogenous estrogen metabolism and individual pharmacokinetic differences has not yet been made but was discussed in detail as theoretically possible.

The major focus of this workshop was on the significance of these individual differences in steroid hormone absorption and metabolism in terms of potential physiologic effects and side effects. Three different contraceptive methods were discussed in detail - oral contraceptives which contain both an estrogen and a progestin, injectable steroid contraceptives and NORPLANT[R] progestin-only implants. The purpose of this commentary is to highlight several issues that will be critical when addressing the question of individual variability as it relates to these three steroid contraceptive methods.

L. DORFLINGER • Research Division, Office of Population, U.S. Agency for International Development, Washington, D.C. 20523-1819. Current Address: Family Health International, P.O. Box 13950, Research Triangle Park, NC 27709.

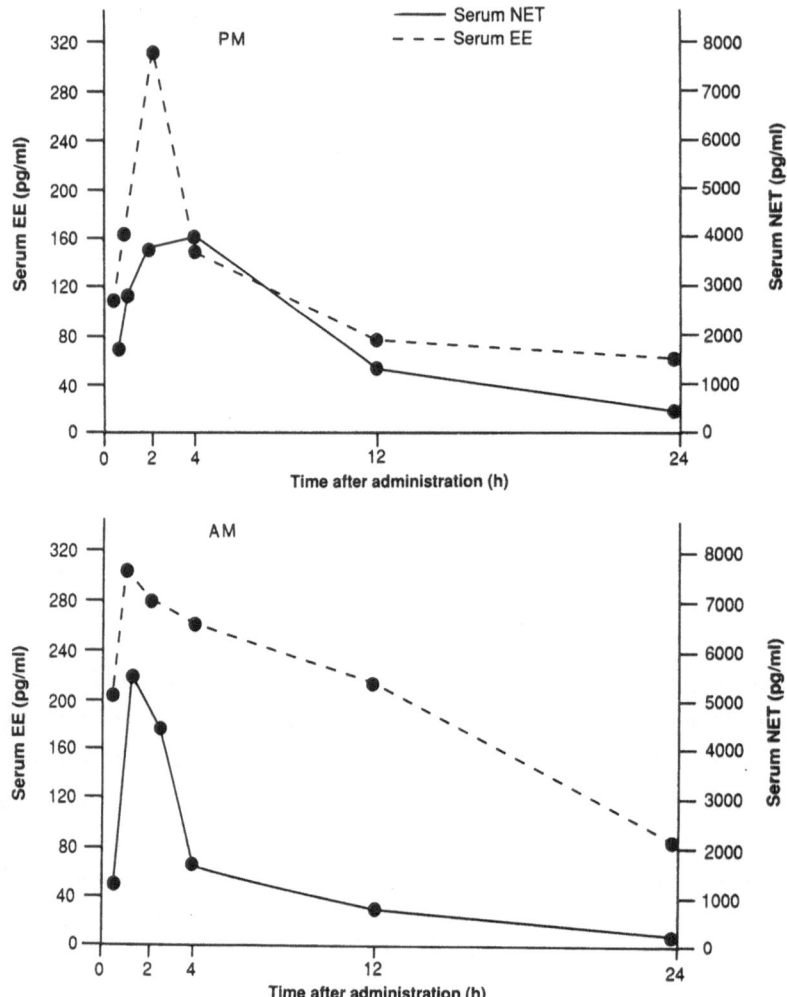

Figure 1. Serum levels of ethinyl estradiol (EE) and norethindrone (NET) for subject #5 after evening (top) or morning (bottom) dose of 50ug EE plus 1mg NET. Data: Kirawat and Fotherby-(1).

2. HORMONE CONCENTRATION RATIOS

Since estrogens and progestins regulate the actions of each other (see below), the question of variability of steroid absorption and metabolism as is relates to physiologic effects of any of the hormonal methods of contraception cannot be adequately addressed without taking into consideration circulating concentrations of both estrogen and progestin. For oral contraceptives, these components will both be synthetic. For the most widely used monthly injectables, a synthetic progestin is given with an ester of estradiol. Finally, for the injectables given every two to three months and for NORPLANT[R], a synthetic progestin is given alone and any circulating estrogen is endogenous.

With few exceptions, investigators who have studied pharmacokinetics of contraceptive steroids have not reported both progestin and estrogen levels for individual women. Rather, group means are reported along with the overall percent of women experiencing a certain side effect. However, if women are evaluated individually, it becomes clear how variable this ratio can be from woman to woman, and within the same woman at different times. For the purposes of demonstrating this point, I have replotted data published by Kirawat and Fotherby (1) to evaluate ratios of estrogen and progestin for individual women (Figures 1 and 2). These authors evaluated serum concentrations of ethinyl estradiol and norethindrone in the same women following early morning and evening administration, so that both intraindividual and interindividual variability are demonstrated in their data. Figures 1 and 2 are examples of two of the six women studied. Values for ethinyl estradiol and norethindrone are plotted in the same figure. These data, as well as the data on the other four women when plotted similarly, demonstrate several things: there is interindividual variation for both hormones; there is intraindividual variability for both hormones; and estrogen and progestin levels appear to vary independently, particularly as seen in Figure 1. Given this variability, one should not be surprised that some women experience steroid-related side effects while others do not.

The importance of the ratios of the two hormones, in addition to the absolute levels of progestin or estrogen following administration of a contraceptive, cannot be overstated. Estrogens and progestins interact at a cellular level by regulating the actions of the other. One way this regulation expresses itself is through modulating receptor levels. Estrogens increase the level of progesterone receptors; progestins decrease both estrogen and progestin receptors (2). Adding to the confusion, in some cell types, recent reports suggest that progestin receptors are not regulated by estrogen (3), and in other cell types progestin does not down-regulate its own receptor (4). Another rarely discussed, but potential confounder, is the fact that synthetic progestins also have some level of androgenic activity which can have several effects. On the one hand, androgens decrease (whereas estrogens increase) the level of SHBG and thereby alter levels of free progestin which bind tightly to this serum protein.

On the other hand, in the breast, androgens have been shown to directly antagonize estrogen-stimulated growth (5). There is one interesting and clear demonstration where changing the estrogen to progestin ration had a negative side effect (ovulation) which was opposite in direction to that originally predicted. The WHO Human Reproduction Programme has two montly injectables which it has developed and is introducing in select countries worldwide. These products are DMPA (25.0 mg) plus estradiol cypionate (5.0 mg), and norethindrone enanthate (50.0 mg) plus estradiol valerate (5.0 mg). During the development process for these two injectable, WHO supported research studies which examined the effect of varying the absolute dose of the two drugs in each preparation as well as the ratio of the progestin to estrogen. The results of one component of this research are shown in Table 1, which describes the percentage of women who ovulated during the third month

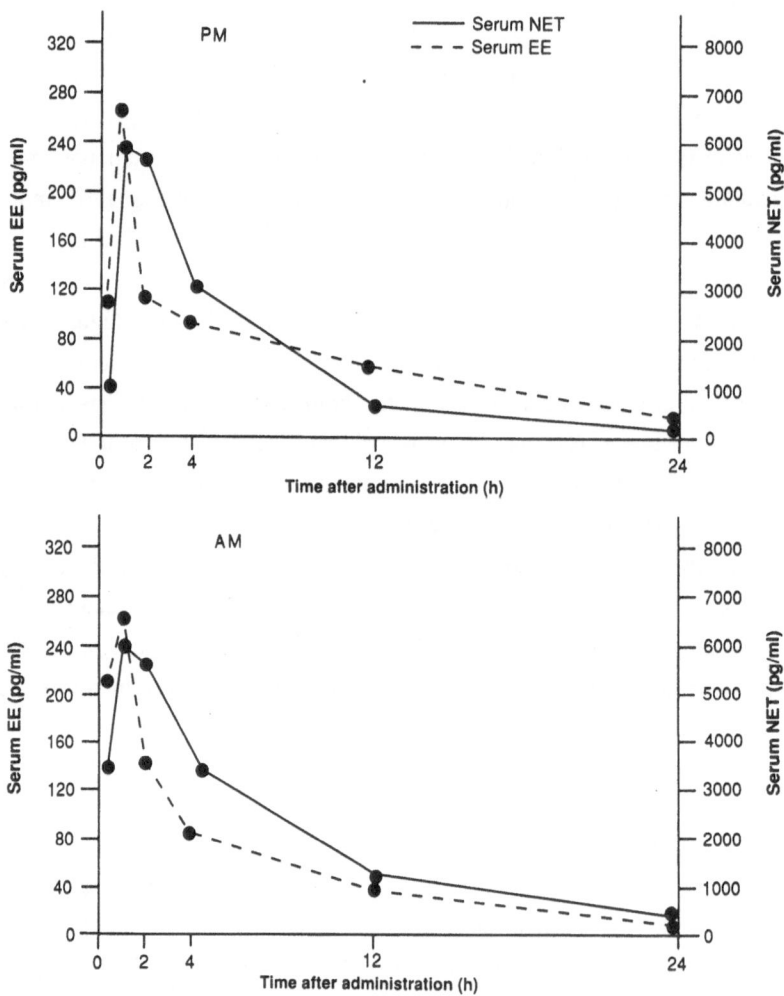

Figure 2. As for Fig. 1, but for subject #1. Data from Kirawat and Fotherby (1).

of treatment, as well as the first and second month following the cessation of treatment. Increasing the estrogen to progestin ratio for both progestins (from 0.2 to 0.4 for the DMPA formulation and from 0.1 to 0.2 for the norethindrone enanthate formulation) resulted in an unexpectedly high rate of ovulation (ovulation being defined as progesterone values above 5 mg/ml in three consecutive serum samples). Since these products do not rely on compliance or variation in absorption from the gut, it would seem reasonable to conclude that the exogenous hormone levels themselves engendered a physiologic response along the hypothalamic-pituitary-ovarian axis that led to ovulation. It then follows that other side effects such as breakthrough bleeding might also be related to these ratios.

3. BIOAVAILABLE OR FREE HORMONE

The hormone ratios illustrated in Figures 1 and 2 are for the total circulating estrogen and progestin. However, the true bioavailability of a hormone is strongly determined by its binding to plasma proteins such as sex hormone binding globulin (SHBG). Estradiol or ethinyl estradiol are only weakly bound to serum proteins; however, synthetic progestins are tightly bound to SHBG such that less than five percent is freely available (6). In a review of the potency of progestins, Fotherby (6) recently concluded that serum concentrations do not correlate with dose or pharmacologic activity. Rather, it is the unbound (free) fraction that is important. Therefore, for oral contraceptives, it would

Table 1. Percentage of Ovulatory Cycles after Administration of DMPA or Net-enanthate-Based One-a-month Injectable Formulations. Data from WHO, reprinted with permission.

FORMULATION	DOSE (mg)	3rd TREATMENT MONTH	1ST FOLLOW-UP MONTH	2ND FU MONTH
DMPA E2-Cyp E/P ratio (n=20)	12.5 2.5 0.2	0	60	90
DMPA E2-Cyp. E/P ratio (n=24)	12.5 5.0 0.4	42	100	100
NET-EN E2-Val E/P ratio (n=23)	25.0 2.5 0.1	0	26	70
NET-EN E2-Val E/P ratio (n=22)	25.0 5.0 0.2	32	73	100

be the ratio of the free ethinyl estradiol to the free level of progestin that is the strongest determinant of the biologic response at an end organ.

The values for free hormones are almost never reported; however, such determinations are critical to the understanding of the link between pharmacokinetic response and physiologic end response. This fact is clearly illustrated in one study of NORPLANT[R] users where Olsson et al. (7) showed that women who conceived while using NORPLANT[R] had a lower "free levonorgestrel index." While the free levonorgestrel index (the ratio of total levonorgestrel to SHBG x 100) is not an exact reflection of the amount of free levonorgestrel because it does not take into consideration other binding proteins or other steroids bound to SHBG, it is a reasonable proxy and is clearly a better reflection of levonorgestrel effects than plasma total levels of this steroid.

Future research that attempts to correlate hormone concentrations with various effects or side effects should therefore use free hormone values rather than total serum concentrations.

4. SUGGESTIONS FOR FUTURE RESEARCH

Any future research which attempts to evaluate individual differences in absorption and metabolism of steroids, and their impact on side effects, will be complicated. Therefore, it is critical to design studies that optimize the chances of linking individual variability and side effects. In light of this, the following two suggestions are made:

A.   Reanalysis of existing data: Available data, for example the data from the WHO trial described above, should be analyzed to determine hormone levels in subjects who experienced a certain side effect e.g. ovulation in the WHO study. Specifically, it should be determined, if possible, from these existing data whether the absolute serum levels of progestin and estradiol (total and/or free) or the relative levels of the hormones were tcorrelated with the side effect of interest.

B.   New prospective study: Select an oral contraceptive formulation that has a well-documented and high incidence of a particular side effect. For example, the lowest dose norethindrone oral contraceptive formulations such as Modicon/Brevicon have a very high rate of breakthrough bleeding when compared to higher dose formulations in clinical trials. In some trials the incidence is as high as 20 to 25%. Women would be recruited and pharmacokinetic analyses (including the determination of free progestin, in addition to total hormone) done at several points in time. Also, other indicators such as preadmission levels of endogenous hormone metabolites in urine and body fat would be determined. Side effects would be carefully monitored over the course of several months of treatment and an attempt made to correlate differences in absolute hormone levels, and ratios of free progestin to estrogen, with resulting side effects. It is important to have objective, verifiable side effect end points in such a study, which is the reason bleeding is proposed for this trial.

References

1. Kiriwat O, Fotherby K. Pharmacokinetics of oral contraceptive steroids after morning or evening administration. *Contraception* 27:153-160. 1983.
2. Clark JH, Schraer WT, O'Malley BW. Mechanisms of Action of Steroid Hormones. In: Wilson JD, Foster DW, eds. *Williams Textbook of Endocrinology.* Philadelphia, PA: WB Saunders Company, 1992: 35-90.
3. Wang S, Counterman LJ, Haslam SZ. Progesterone action in normal mouse mammary gland. *Endocrinology* 127:2183-2189. 1987.
4. Okulicz WC, *et. al.* Biochemical and immunohistochemical analyses of estrogen and progesterone receptors in the rhesus monkey uterus during the proliferative and secretory phases of artificial menstrual cycles. *Fertility Sterility* 53:913-920. 1990.
5. Simard J, *et. al.* Regulation of progesterone-binding breast cyst protein GCDFP-24 secretion by estrogens and androgens in human breast cancer cells: A new marker of steroid action in breast cancer. *Endocrinology* 126:3223-3231. 1990.
6. Fotherby K. Potency and pharmacokinetics of gestagens. *Contraception* 41:533-550. 1990.
7. Olsson S-E, *et. al.* Plasma levels of levonorgestrel and free levonorgestrel index in women using NORPLANT[R] implants or two covered rods (NORPLANT[R]-2). *Contraception* 38:227-234. 1987.

EXPERIMENTAL DESIGN FOR FUTURE RESEARCH

James Trussell[1]

The discussion during the past two days has revealed that the value of future research might be considerably advanced if only a small amount of additional information were routinely collected. The cost of acquiring such information varies, but in many cases appears to be quite modest. The previous discussion distinguished three types of studies -- clinical trials of contraceptives, laboratory bioavailability studies, and population-based surveys -- so my remarks are grouped accordingly.

1. CLINICAL TRIALS

Information on body composition and topography could be collected at minimal additional cost through bio-impedance analysis (BIA) and measurement of hip and waist size. Such information could then be related to perceived side effects and actual discontinuation through the use of standard statistical techniques (multivariate life tables or hazard models).[1,2] Information on diet, in contrast, would be both very difficult and expensive to collect at the individual level. Nevertheless, knowledge of the impact of diet on side effects and discontinuation could easily be gained by stratification across groups with large average differences in diet. One could, for example, select two groups of Han Chinese -- one which has a diet relatively rich in fat and another which has a diet with high fiber content. In a similar way one could examine two groups of Indian women, one consisting of vegans and the other with a diet with high fat content.

J. TRUSSELL • Office of Population Research, Princeton University, Princeton, NJ 08544

Steroid Contraceptives and Women's Response,
Edited by R. Snow and P. Hall, Plenum Press, New York, 1994

Knowledge about the biological basis of perceived side effects could be gained by selecting a group of women who were protected from the risk of pregnancy because their partners had a vasectomy and randomly assigning them to either a hormonal contraceptive or a placebo. Clinicians would then have a firmer basis on which to distinguish true biological effects of synthetic hormones from a placebo effect. The extent to which side effects are related to perceived risk of pregnancy could be judged by comparing the results with those from a third group of women who relied on that hormonal contraceptive for pregnancy prevention. This knowledge could be very useful in counseling, provided that it was not employed in a coercive or threatening manner to intimidate clients or dismiss in an insensitive manner their real concerns.

Our understanding of risk-taking behavior and of imperfect use of contraception, as well as the true efficacy of methods other than hormonal contraceptives, is severely limited by the fact that information on perfect or imperfect use is not routinely collected on a cycle-by-cycle basis.[3,4] Such an understanding would be especially valuable when counseling couples and is necessary for truly informed consent. Hence, there is much to be gained from routinely collecting information on imperfect use during each cycle.

Given the obvious role that coital frequency plays in the efficacy of a particular contraceptive for a given couple, it is regrettable that coital behavior is so rarely examined. In part, this omission is due to the fact that the coital frequency varies over time, so that information must be collected on a cycle-by-cycle basis to be of much use. Moreover, retrospectively reported data are suspect, even if the recall period is only one cycle.[5] Daily coital logs appear to provide the most reliable information, but these impose a significant burden on respondents and are intrusive enough that behavior might be modified as a consequence of having to record the data. Thus, while there is consensus on the desirability of collecting information on coital frequency -- to inform our understanding of both contraceptive efficacy and sexual behavior more generally -- there remains the considerable methodological challenge of designing ways to collect information reliably but unobtrusively.

Finally, the usefulness of clinical trials would be greatly enhanced if data were routinely documented well and made freely available to other investigators upon request. Indeed, such expectations should be built into the philosophy and the budget of contraceptive trials.

2. BIOAVAILABILITY STUDIES

Several investigators warned throughout the conference that the bioavailability of contraceptive hormones displays variability both across and within individuals, so that the typically small samples (Ns in the single digits) characteristic of bioavailability studies preclude identification of the importance of hypothesized sources of systematic effects. The challenge is to disentangle the effects of body composition and topography, diet, other drugs, age, and other factors such as race on the parameters of bioavailability (e.g. C_{max}, T_{max}, $T_{1/2}$, and AOC).

Given this large list of potential candidates for important sources of variability, it is clear that much larger studies would be needed. It is equally clear that for results of different studies to be comparable, far more care must be exercised to ensure that the same protocols are followed. Rather than trying to answer all questions with a few large studies, a more sensible strategy might be to examine with smaller studies (say $N=25$) the importance of single factors by holding other factors constant. The danger of such an approach is obvious, since it precludes identification of important interactions and confounding effects, but it does have the twin advantages of being able to spot truly large effects (called "slam-bang" effects in the literature on evaluating interventions) and of being able to be conducted in this era of dwindling research funds.

It is also clear that intrapersonal variation, while an important subject in its own right, is irrelevant to the identification of the factors considered above, since these do not change much, if at all, within an individual over a short time span. Hence, if the object is to answer the questions raised above, a study of thirty individuals with one observation per subject is far superior to a study of ten individuals with three observations per subject.

3. POPULATION SURVEYS

Much was said at this meeting about contraceptive choice. While the word *choice* is indeed a desirable goal, in that it implies informed consent to use a particular contraceptive from a rich array of options, it is clear that use of a contraceptive often does not really reflect a choice. In many circumstances, options are severely curtailed by the actions of governments and individual providers. Nevertheless, even in cases in which access is relatively unrestricted, we have only minimal understanding of why women or couples choose the methods they do use.

Statistical modelling of contraceptive choice is difficult for several reasons. First, there are many potential options. Second, contraceptive use is a dynamic process, whereby current use is surely influenced by one's past experience, by one's evolving demands of attributes of a contraceptive (convenience, efficacy at preventing pregnancy, efficacy at reducing STD transmission, safety, cost), and by evolving perceptions (both accurate and inaccurate) about the attributes of each option.

Given this perspective, it is clear that statistical models of the determinants of method choice that do not incorporate previous history are unlikely to be useful. It is also clear that to study contraceptive choice, the investigator must employ a sample that is population - based; not only are the subjects in clinical trials self-selected, but such trials (and indeed most family planning providers) do not offer the full range of options actually available to the population. However, actually estimating a reasonably specified model of choice, even if the data were available, would be a formidable challenge since the potential number of important factors and the potential number of transitions is so vast.

Moreover, the fact that we do not understand well (in a quantitative sense) the factors that do influence choice has important implications for our understanding of both efficacy and side effects. The fact that people choose their contraceptive methods means that all samples of users are self-selected. Hence, conclusions about efficacy and side effects are very difficult to generalize to any well-defined population. For example, empirical studies reveal that oral contraceptives are far superior to spermicides in preventing pregnancy.[7] But because this information is widely available, it may well be the case that those who choose spermicides are not highly motivated to prevent pregnancy. Indeed, it is plausible that if women currently using oral contraception were forced to use spermicides instead, they would experience far lower failure rates than do those who currently choose to use spermicides.

Explicitly modelling such unobserved heterogeneity is in principle easy, but the difficulty is that interpreting the results behaviorally is quite problematic. Infinitely many different behavioral models yield identical predictions in the space of observable outcomes, so that identification is achieved solely by an assumed functional form.[8] Published attempts to model both selection and efficacy do not inspire confidence. For example, one study concluded confidently that oral contraceptives have higher efficacy than sterilization.[9] Hence, while I am skeptical that contraceptive choice can be modelled well, I am equally convinced that more qualitative research from the user's perspective can enhance our understanding of the factors users deem important.

References

1.   Cox D, Oakes D. *Analysis of Survival Data*. Chapman and Hall, London. 1984.
2.   Kalbfleisch J, Prentice R. *The Statistical Analysis of Failure Time Data*. John Wiley, New York. 1980.
3.   Trussell J, *et. al.* Contraceptive failure in the United States: An update. *Studies in Family Planning* 21(1):51-54. 1990.
4.   Trussell J, Grummer-Strawn L. Contraceptive failure of the ovulation method of periodic abstinence. *Family Planning Perspectives* 22(2):65-75 and *International Family Planning Perspectives* 16(1):5-15. 1990.
5.   Hornsby PP, Wilcox AJ. Validity of questionnaire information on frequency of coitus. *American Journal of Epidemiology* 130(1):94-99. 1989.
6.   Blanc AK, Rutenberg N. Coitus and contraception: the utility of data on sexual intercourse for family planning programs. *Studies in Family Planning* 22(3):162-176. 1991.
7.   Trussell J, Kost K. Contraceptive failure in the United States: A critical review of the literature. *Studies in Family Planning* 18(5):237-283. 1987.
8.   Trussell J, Rodríguez G. Heterogeneity in Demographic Research. In Julian Adams, David Lam, Albert Hermalin and Peter Smouse (eds.), *Convergent Questions in Genetics and Demography*. Oxford University Press, New York. 1990.
9.   Rosenzweig MR, Schultz TP. Schooling, information, and nonmarket productivity: contraceptive use and its effectiveness. *International Economic Review* 30(2):457-477. 1989.

LOOKING FORWARD: NEW RESEARCH PRIORITIES

Lincoln Chen[1]

Thirty people, three days, at the lovely Exeter Inn. I calculated that that's 90 person-days, which is a quarter of a person-year. With all the preparation that went into the workshop at least one person-year of effort has been invested in this important enterprise. Our intent of selecting the Exeter Inn in New Hampshire was to enable us to catch some of the lovely fall colors; but it rained! Fortunately, it's now clearing up, and so on the way to the airport, many of us will be able to see the changing fall colors of New England.

The workshop *Steroid Contraception and Women's Response* was an ambitious intellectual journey with many important landmarks. It pursued a sandwich strategy, beginning first with biosocial dimensions of selection and side effects of contraceptive methods, then moving into the biologic variability and pharmacokinetics of hormonal methods, and concluding with policy, program and research implications. Through it all, three shifting themes among the participants were noteworthy.

This was firstly a multi-disciplinary workshop for exchange, dialogue, and research planning between biomedical, social, and policy-oriented scientists. In much of the discussion, we were communicating to our disciplinary peers, despite the presence of colleagues from allied fields. There were, however, many successful "leakages," (fortunately) across much of the exchange between affinity groups. To my mind, the individual papers benefited enormously from this cross-disciplinary exchange.

A second underlying premise was that we were not dealing with the development of new contraceptive methods. Rather, the focus of this workshop was aimed at improving the impact, health and otherwise, of existing methods that have already been developed and are

L. CHEN • Department of Population and International Health, Harvard School of Public Health, 665 Huntington Avenue, Boston, MA 02115

Steroid Contraceptives and Women's Response,
Edited by R. Snow and P. Hall, Plenum Press, New York, 1994

now in use. That very fundamental focus, I suggest, profoundly shaped some of the exchange, even misconceptions, about the importance of issues such as biologic variability and side-effects.

Thirdly, we too often spoke past each other in terms of the objectives of steroid contraception. All of us agreed that steroid contraception is an important technology for birth control. But why birth control? demographic? health? human rights? These different underlying "whys" generated varying expectations of the technology and consequently of our exchange.

Let me summarize the workshop from the perspective of four simple questions.

(1) Is there biologic variability in steroidal metabolism between human populations based upon empirical pharmacokinetic data? (2) And if so, what are the determinants or correlates of these variabilities? We devoted the bulk of this workshop to these two questions, but we've also wrestled with third and fourth questions. (3) Does this variability affect the successful use of contraception? (4) And, if it does, so what? What should we do?

Let me deal with these questions briefly. The answer to the first question, (Is there biologic variability?), is clearly yes. The differences between Thai, Mexican and Chinese women presented by Yasmin Hemayet Ahmed, Josue Garza-Flores and Sang Guo-wei, respectively, are impressive. Some of the differences according to Kenneth Fotherby and Joseph Goldzieher are due to pharmacokinetics in terms of dosage amount, the route of administration, and product formulation. We even learned that the shape and size of *one's behind* could affect these parameters! But also physiologic variabilities, absorption, the first passage, enterohepatic circulation, and sex-hormone-binding globulins are important. We learned from Soledad Diaz that because of its metabolic role, the liver is really a "reproductive organ," a manager of steroidal pharmacokinetic traffic.

Our papers did not address the frontiers of the biologic sciences, which is what is happening at the receptor site, or, as Sioban Harlow noted, what is happening at the mitochondrial level in target organs? These binding and sub-cellular mechanisms are the net effect of much of this metabolic traffic.

There were solid hypotheses regarding the correlates and determinants of pharmacokinetic variability, the most noteworthy probably being body weight, adipose tissue, its distribution and type, its variable functions, and question of diet. Nutrition and diet apparently exert a profound effect on steroidal absorption and excretion, and perhaps metabolism as well. We also learned from David Back the question of drug metabolism and smoking, and Joan Kaufman asked questions about illness and disease. So, many good hypotheses were advanced and examined.

How do all these pharmacokinetic data, biologic variability, and their correlates relate to the consequences of contraceptive use? In this relationship, we entered slippery ground. Different views were expressed. Harold Nash commented that none of these findings had any relationship to the development of specific contraceptive methods. Contraceptive technology is unaffected by these dynamics, he noted. However, if one examines these methods as they are used and as they relate to diverse communities and individuals, then the workshop was not simply dealing with method development, but with a technology in use within specific social and programmatic contexts. The workshop approach, therefore, was from the "user perspective," integrating biological, social, and programmatic dimensions influencing contraceptive use amongst diverse communities around the world. In this approach, contraception is viewed as a tool to achieve social objectives, fulfilling multiple objectives -- birth control for population control, health, and human rights. Biologic efficacy of the technology is critical, but so too are side-effects, health consequences, preferences, convenience, cost and other important attributes of the technology.

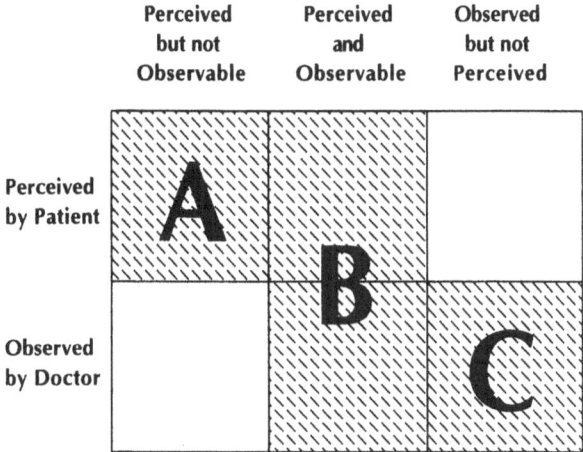

Figure 1. Side Effects of Steroid Contraception.

Biologic variability, it is postulated, influences many of these attributes, especially the entire issue of side effects. Side effects are experienced not in the laboratory, but by people who are using this tool. The traditional view of contraceptive development and ultimate use is a top-down process involving a message (contraception), a messenger (service provider), and an audience (client). How often are we in danger of shooting the messenger who brings a questionable message (the message being the contraceptive method and the messenger being a health worker)? Conversely, messengers might have difficulties without the right message. As Peter Hall noted, we ourselves need to learn how to communicate the message (the technology) much better in terms of meeting the needs of the audience.

Two strong and apparently polar views were expressed about side effects. One view was that side effects are objective, observable phenomena that can be measured by control studies which remove placebo effects. This approach is based upon the natural   sciences which have proven to be of enormous value to clinicians. A second view was that this "objectivity" is not the sole issue. Whatever is perceived or experienced by the client using a technology is legitimate, irrespective of underlying placebo effects.

Are these two views compatible? This is one fundamental question of the workshop, in terms of theory, measurement, and ultimately technology usefulness.

In Figure 1, a framework for undertaking the fundamentally different phenomena of perceptual and objective assessment of side-effects is shown. A future research challenge is to seek the intersection and the harmony between these two different approaches to side-effects. The work presented in Figure 2 comes from a multi-disciplinary group at Harvard where we have been studying population and health transitions, including mortality and morbidity change. Among those involved are Chris Murray (health economist), Amartya Sen (economist), Arthur Kleinman, (anthropologist), in addition to myself.

Side effects, like all morbidities, can be of two types. Some can only be self-perceived (and reported) by an individual; others can only be assessed by an external observer; some side effects are both self-perceived, and observed externally.

In the first category are some morbidities, such as headache or fatigue, which are entirely perceived by an individual. These cannot be objectively observed and measured. They are entirely self-perceived side effects. There is no objective, observable way of measuring an individual's pain and suffering.

In the second category are some side effects that may not be perceived at all; they can be, however, readily measured and assessed by an external observer. An example is hypertension or high blood pressure. Hypertension as a side-effect of contraception would be an externally observable side-effect, but not self-perceived. It can only be observed. A client does not perceive hypertension. Another example is a metabolic malfunction, for example, disturbances of insulin and blood sugar metabolism associated with diabetes.

So there are two categories where, in one case the client perceives the side-effect but an external observer cannot measure it; and there's another morbidity where an external observer can measure the phenomenon, but the client cannot perceive it.

There are also side-effects which are both perceived and externally observed. One example is Elizabeth Belsey's data on menstrual bleeding patterns. Bleeding is something that both the client can perceive and can also be presumably externally measured (under appropriate circumstances). So, too, would be a skin rash.

The key point of these three types of side-effects is both the client as well as an external observation are needed to assess the full constellation of side effects. One cannot assess side effects from either approach alone. This understanding can constitute a framework on how to move the different approaches about side-effects together.

This suggests that both a scientific case-control estimate of side-effects (eliminating the background noise of placebos) as well as self-perceived side-effects (irrespective of the level of placebo effects) are important. On the one hand, it would be foolhardy to falsely attribute placebo effects to a new technology. On the other hand, it would be unwise to claim as irrelevant a method with high self-perceived side-effects, even if placebo effects are high. On the later point, take as an example, a method that causes 100 percent of users to report self-perceived side-effects. If placebo effects were 60 percent, one would conclude that only 40 percent of the side-effects can be attributed to the method *per se*. Yet, from the client perspective, everyone would be experiencing side-effects (irrespective of the placebo effect). This would hardly be a testimony to the acceptability of this method.

The thrust of this analysis is that both self-perceived as well and externally observed side-effects are legitimate. It is not necessary that these different aspects of morbidity be congruent, although in some cases they reach across to each other. In terms of future research, we should examine the "two-way street" of self-perceived and externally observed side-effects. Such will require strengthening the long linkage between the laboratory and the community. Study designs are very much at heart of this linkage, as most of them are constructed from a lab-to-field paradigm. The classical design of clinical trials that Peter Hall described is illustrative. Other design paradigms are also needed that move the enquiry from the client back into the laboratories, as Rachel Snow suggested.

In the future, four research approaches, I believe, deserve priority. First is fuller analysis of existing data. Carlos Huezo and Peter Hall noted that there are very rich data sets that deserve analytical attention to answer high-priority questions. The second is to strengthen the intellectual basis of many ongoing studies that can be improved with sounder theoretical frameworks. A third approach would be opportunities to add on key components to existing studies or planned studies. Some add-on variables include body composition, pharmacokinetic, or side-effect components. Finally, much of what is needed would require entirely new study designs, and much stronger participation of the social sciences in managing the research agenda.

As we move forward, we're going to have to cope, not only with biologic variability, but also with social, cultural, side-effect, and geographic variability. In other words, it may be legitimate to find side-effects of 100 percent in one place, but only 5 percent in another place, simply because of differing peoples and ecologies. To meet this challenge will call for much stronger capacity for research by local participants in the various countries in which contraception is being used.

Let me conclude by quoting from the Sunday, October 14th *Boston Globe* a commentary by Anita Diamond, who wrote on the oral pill, which had its 30th birthday this year.

"Thirty is an awesome milestone. Thirty is all grown up but also somehow very young. In 1960 the Food and Drug Administration approved a license for a drug that effectively and reliably prevented conception. You could practically feel the earth move. Today 60 million women use birth control pills. The pill has become one of the most widely prescribed drugs in history. Eighty percent of American women born between 1945 and 1964 have spent some portion of their lives on the pill. While our notion of sex, family and gender roles might have change without it, the pill was a fundamental catalyst in the continuing social revolution of our times. Birth control pills gave its users much more than reliable control over when and how to bear children. The pill widened the road that stretched before a generation of young women. Most of all, the pill transformed sex by separating the act of love making from the act of baby making."

This commentary related mostly, of course, to American women, not to women of the developing world. But its conclusion is relevant to all peoples: "In the grand scheme of things, 30 years is the blink of an eye. In the course of human events, 30 is a span of a generation. In a lifetime, 30 marks the beginning of maturity. After three decades, the pill is a radical innovation, a familiar friend, a challenge to our collective conscience. Many happy returns."

CONTRIBUTORS

Yasmin H. Ahmed
Bangladesh Institute of Research for Promotion of Essential and Reproductive
Health and Technologies (BIRPERHT)
25 Shyamoli, Mirpur Road
Dhaka 1207
Bangladesh

David Back Ph.D.
The University of Liverpool
Department of Pharmacology and Therapeutics
New Medical Building
Ashton Street
Liverpool L69 3BX, United Kingdom

Elizabeth Belsey Ph.D.
Global Program on AIDS
World Health Organization
1211 Geneva 27, Switzerland

Leon Bradlow Ph.D.
Strang Cornell Cancer Research Laboratory
510 East 73rd Street
New York, New York 10021-4004

Judith Bruce
Population Council
1 Dag Hammarskjold Plaza
New York, New York 10017

Lincoln Chen M.D., M.P.H.
>   Department of Population and International Health
>   School of Public Health
>   665 Huntington Avenue
>   Boston, Massachusetts 02115

Soledad Diaz M.D.
>   Instituto Chileno Midicina Reproductiva
>   Jose Victorino Lastarria 29 Depto. 101
>   Casilla 6006, Santiago, Chile

Laneta Dorflinger Ph.D.
>   Family Health International
>   P.O. Box 13950
>   Research Triangle Park, North Carolina 27709

Kenneth Fotherby Ph.D.
>   Royal Postgraduate Medical School
>   University of London
>   London W12 ONN, United Kingdom

Josue Garza-Flores M.D.
>   Instituto Nacional de la Nutricion Salvador Zubiran
>   Departamento de Biologia de la Reproduccion
>   Col. Tlalpan Vasco de Quiroga
>   14000 Mexico, D.F.

Barry Goldin Ph.D.
>   Tufts University School of Medicine
>   136 Harrison Avenue
>   Boston, Massachusetts 02111

Joseph Goldzieher M.D.
>   Texas Tech University
>   1400 Wallace Boulevard
>   Amarillo, Texas 79106

Gary S. Grubb M.D., M.P.H.
>   Family Health International
>   P.O. Box 13950
>   Research Triangle Park, North Carolina 27709

Peter Hall
>   World Health Oganization
>   1211 Geneva 27, Switzerland

Siobhan Harlow Ph.D.
>   Departamento de Epidemiologia
>   Escuela de Salud Publica de Mexico
>   Instituto Nacional de Salud Publica
>   Cuernavaca, Morelos, Mexico

Michael Humpel Ph.D.
>   Schering AG
>   Muellerstrasse 178
>   13342 Berlin
>   Federal Republic of Germany

Joan Kaufman Sc.D.
>   Department of Population and International Health
>   Harvard School of Public Health
>   665 Huntington Avenue
>   Boston, Massachusetts 02115

Christopher Longcope M.D.
>   Department of OB/GYN
>   UMASS Medical School
>   Worcester, Massachusetts 01655

Harold Nash Ph.D.
>   The Population Council
>   Center for Biomedical Research
>   1230 York Avenue
>   New York, New York 10021

Carla Obermeyer Sc.D.
>   Department of Population and International Health
>   Harvard School of Public Health
>   665 Huntington Avenue
>   Boston, Massachusetts 02115

Michael Orme M.D.
>   The University of Liverpool

Department of Pharmacology and Therapeutics
New Medical Building
Ashton Street
Liverpool L69 3BX, United Kingdom

Dale Robertson M.D.
The Population Council
Center for Biomedical Research
1230 York Avenue
New York, New York 10021

Guo-wei Sang M.D.
Family Planning Research Institute of Zhejiang
Zhejiang Academy of Medical Sciences
Hangzhou, Zhejiang 310013, People's Republic of China

Pramilla Senanyake Ph.D.
International Planned Parenthood Federation
Regent's College
Inner Circle, Regent's Park
London NW1 4NS, United Kingdom

Rachel Snow Sc.D.
Department of Population and International Health
Harvard School of Public Health
Boston, Massachusetts 02115

James Trussell Ph.D.
Office of Population Research
Princeton University
Princeton, New Jersey 08544

Lucy E. Wilson M.S.
University of Maryland School of Medicine
700 Portland Street
Baltimore, Maryland 21230

INDEX